MODERN EARTH SCIENCE

Section Reviews with Answer Key

HOLT, RINEHART AND WINSTON
Harcourt Brace & Company
Austin • New York • Orlando • Atlanta • San Francisco • Boston • Dallas • Toronto • London

Printed in the United States of America

ISBN 0-03-051439-8 345 021 00 99 98

Contents

About SECTION REVIEWS

Modern Earth Science is divided into 91 sections. A SECTION REVIEW blackline master is provided for each section in *Modern Earth Science.* The review pages can be reproduced and distributed to your students. The review pages enable students to test their understanding of each section's objectives.

Each SECTION REVIEW can be used to reinforce or review concepts introduced in the text. In this respect, *Section Reviews* can be used as brief written classroom activities, or they can be assigned as homework. The *Section Reviews* can also be used as quizzes to help evaluate student comprehension of section content.

Each SECTION REVIEW consists of 10 questions. All of the review questions have been correlated to the Section Objectives.

Each SECTION REVIEW is formatted with two types of questions. Combinations of the following types of questions are included:

- True-False
- Multiple-Choice
- Fill-in-the-Blanks
- Short-Answer
- Essay

Answer Keys are provided for the teacher. The correlations to the section objectives can be found in the answer keys.

Name ______________________ Class ______________ Date ______________

Section 1.1

What Is Earth Science?

Read each statement below. If the statement is true, write *T* in the space provided. If the statement is false, write *F* in the space provided.

_____ **1.** Astronomy is a branch of earth science.

_____ **2.** The atmosphere is one part of the geosphere.

_____ **3.** Earth scientists primarily study the geosphere.

_____ **4.** An ecosystem is a largely self-supporting, physically distinct system.

_____ **5.** Biodegradable wastes pose a major threat to the environment.

_____ **6.** Exploration for oil deposits would most likely be done by an ecologist.

Choose the one best response. Write the letter of that choice in the space provided.

_____ **7.** The ozone in the upper atmosphere is a form of:

a. carbon. **b.** hydrogen. **c.** oxygen. **d.** silicon.

_____ **8.** The largest ecosystem is called the:

a. biosphere. **b.** atmosphere. **c.** hydrosphere. **d.** geosphere.

_____ **9.** Which of the following would an astronomer most likely study?

a. a comet's path through the solar system
b. the height of tides during a full moon
c. an iridium-laden rock layer produced by meteorites
d. the amount of solar energy absorbed by a tree

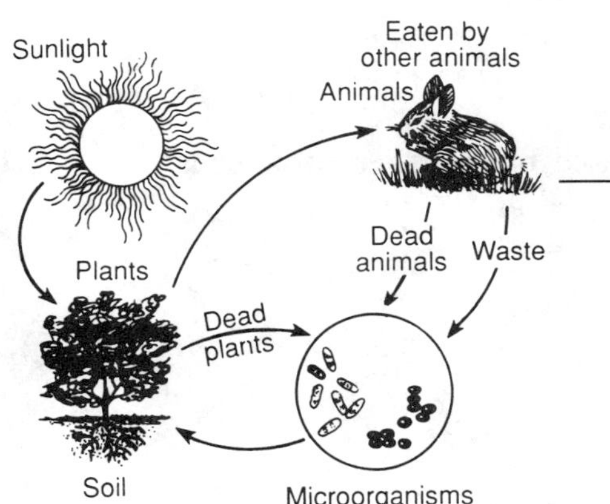

_____ **10.** The relationships shown in this diagram would most likely be studied by:

a. a geologist. **b.** an ecologist.
c. a meteorologist. **d.** an astronomer.

Name ______________________ Class ______________ Date ______________

MODERN EARTH SCIENCE

Section 1.2

Paths to Discovery: Scientific Methods

Read each statement below. If the statement is true, write *T* in the space provided. If the statement is false, write *F* in the space provided.

______ **1.** Dust raised by the impact of a meteorite on Earth may have blocked the sun's rays for many years.

______ **2.** A hypothesis must be tested before any observations can be made.

______ **3.** A kilogram is a standard unit of measurement.

______ **4.** Controlled experiments have confirmed the meteorite-impact hypothesis.

______ **5.** Scientific methods are guides to scientific problem solving.

Complete each statement by writing the correct term or phrase in the space provided.

6. Scientists believe that the iridium found in some rock layers may originally have come from a

______________________ .

7. Comparing some aspect of an object with a standard unit is called

______________________ .

8. A controlled experiment is designed to test a single factor called a

______________________ .

9. Scientists proposed the meteorite-impact hypothesis to explain the extinction of the

______________________ .

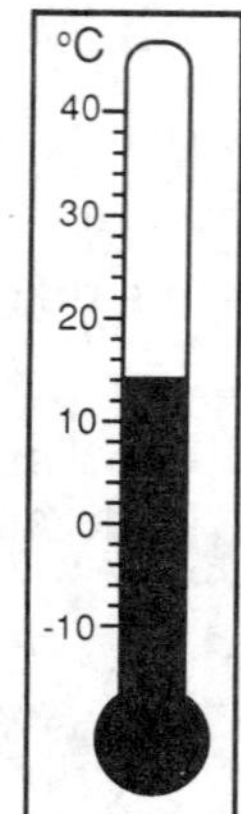

10. The temperature indicated by this thermometer is

______________________ .

Section 1.3

Birth of a Theory: The Big Bang

Complete each statement by writing the correct term or phrase in the space provided.

1. A theory that is well established through research and experimentation is most likely to become a scientific ______________________ .

2. When light passes through a glass prism, it produces a band of colors called the ______________________ .

3. The apparent shift in the wavelengths of energy emitted by an energy source moving away from or toward an observer is known as the ______________________ .

4. When heated, each element produces a series of thin colored lines called a ______________________ .

5. Scientists studying starlight determine what elements are present in the stars by using a ______________________ .

6. The discovery of a red shift in the spectra of galaxies provided evidence that the universe is ______________________ .

Choose the one best response. Write the letter of that choice in the space provided.

_____ **7.** Background radiation evenly distributed throughout the universe likely resulted from:

a. the Doppler effect. **b.** moving galaxies.
c. the Big Bang. **d.** starlight spectra.

_____ **8.** When a light source moves toward an observer, the light waves appear to be:

a. moving slower. **b.** longer.
c. moving faster. **d.** shorter.

_____ **9.** It is thought that before the Big Bang all the matter and energy in the universe was in the form of one:

a. extremely small volume. **b.** solar system.
b. expanding cloud. **d.** galaxy.

Traveling Light Wave

_____ **10.** Look at the diagram. One wavelength is equal to the distance between points:

a. 1 and 3. **b.** 1 and 4.
c. 2 and 3. **d.** 2 and 4.

Name ______________________ Class ______________ Date ______________

Section 2.1

Earth: A Unique Planet

Read each statement below. If the statement is true, write *T* in the space provided. If the statement is false, write *F* in the space provided.

_____ **1.** The outer core is the only layer of the earth that is believed to be liquid.

_____ **2.** When an earthquake occurs, S waves can be detected on the side of the earth opposite the earthquake.

_____ **3.** Because the sun contains large amounts of iron, it has a magnetic field.

_____ **4.** The earth's magnetic field affects an area that extends beyond the atmosphere.

_____ **5.** According to Newton's law of gravitation, the force of attraction between two objects depends on their masses and the distance between them.

Choose the one best response. Write the letter of that choice in the space provided.

_____ **6.** The part of the earth that is solid but has the ability to flow is the:

a. lithosphere. **b.** oceanic crust.
c. asthenosphere. **d.** inner core.

_____ **7.** The earth's atmosphere is made up of mostly oxygen and:

a. nitrogen. **b.** water vapor.
c. hydrogen. **d.** carbon dioxide.

_____ **8.** Seismic waves increase in speed when they enter:

a. the hydrosphere. **b.** a strong magnetic field.
c. the outer core. **d.** a more rigid material.

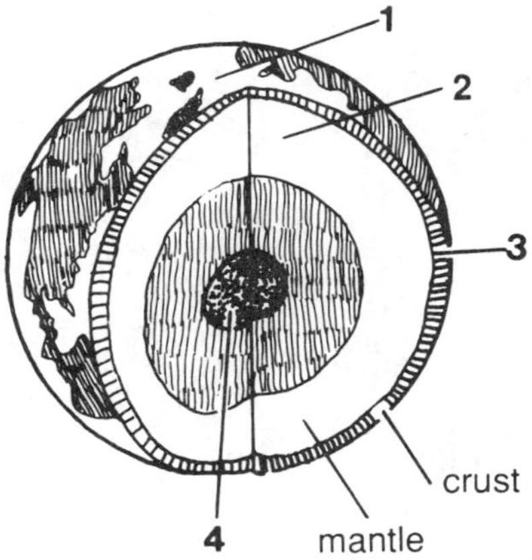

_____ **9.** At which location in the diagram is the Moho found?

a. 1 **b.** 2
c. 3 **d.** 4

_____ **10.** The mass of an object is defined by:

a. how much the object weighs on earth.
b. the amount of water the object displaces.
c. the strength of the pull of gravity on the object.
d. the amount of matter in the object.

Name ______________________ Class ______________ Date ______________

MODERN EARTH SCIENCE

Section 2.2

Movement of the Earth

Read each statement below. If the statement is true, write *T* in the space provided. If the statement is false, write *F* in the space provided.

______ **1.** At its closest point to the sun, the earth is said to be at aphelion.

______ **2.** Precession is the wobbling of the earth's axis as it turns in space.

______ **3.** There are exactly 24 standard time zones around the world.

______ **4.** As the earth orbits around the sun, the direction of the tilt of its axis changes.

______ **5.** Each rotation of the earth takes about one year.

______ **6.** When it is 5:00 P.M. Friday immediately west of the international date line, it is 5:00 P.M. Thursday immediately east of the line.

Choose the one best response. Write the letter of that choice in the space provided.

______ **7.** Approximately how many degrees of the earth's total circumference does each time zone cover?

a. 1° **b.** 15° **c.** 45° **d.** 90°

______ **8.** Each year, the area south of the Antarctic Circle experiences 24 hours of darkness on:

a. December 21 or 22.
b. March 21 or 22.
c. June 21 or 22.
d. September 22 or 23.

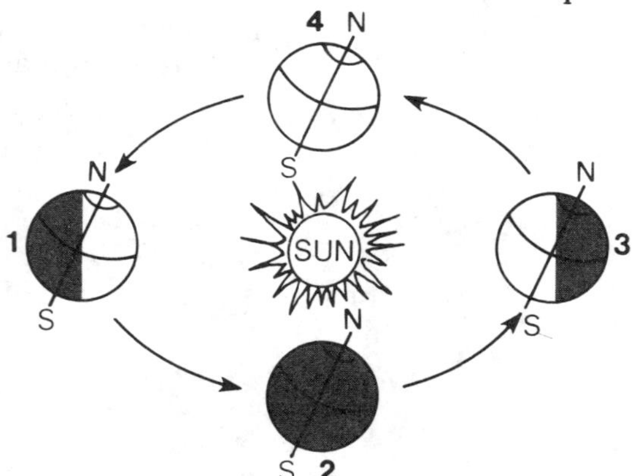

______ **9.** Which number on the diagram represents the earth's position on the autumnal equinox?

a. 1
b. 2
c. 3
d. 4

______ **10.** On the day of the summer solstice, the sun's vertical rays strike the earth along the:

a. Antarctic Circle.
b. Tropic of Capricorn.
c. Arctic Circle.
d. Tropic of Cancer.

Name ______ Class ______ Date ______

MODERN EARTH SCIENCE

Section 2.3

Artificial Satellites

Read each statement below. If the statement is true, write *T* in the space provided. If the statement is false, write *F* in the space provided.

_____ **1.** Any object in orbit around another body with a smaller mass is a satellite.

_____ **2.** *Landsat* satellites can locate forested areas on the earth's surface.

_____ **3.** A satellite encounters more air resistance at higher-altitude orbits.

_____ **4.** The space shuttle is a satellite that can return to the earth's surface.

_____ **5.** A satellite in orbit travels slowest at perigee and fastest at apogee.

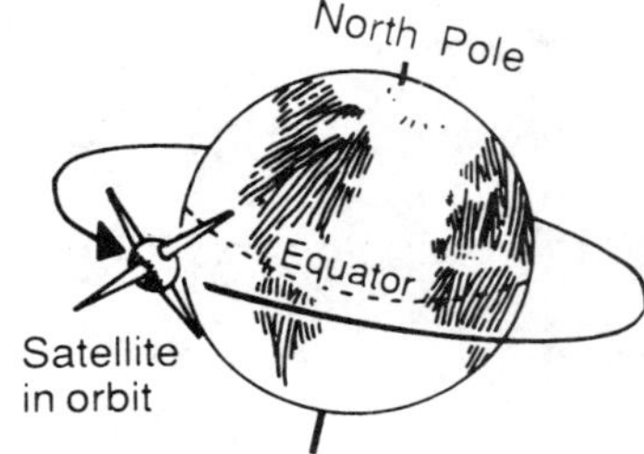

_____ **6.** The satellite in this diagram is traveling in a polar orbit.

Choose the one best response. Write the letter of that choice in the space provided.

_____ **7.** A satellite will remain in orbit if its speed is adequate for its:

a. mass.
b. distance from apogee.
c. altitude.
d. distance from perigee.

_____ **8.** Which satellite network has been used for accurate navigation of ships and aircraft?

a. space shuttles
b. GIS
c. *Landsat*
d. GPS

_____ **9.** Which of the following statements is true of a satellite in a polar orbit?

a. It remains at the same point above the equator.
b. It moves in the same direction as the earth's rotation.
c. It completes one revolution every 24 hours.
d. It passes over a different portion of the earth during each revolution.

_____ **10.** A satellite in a geosynchronous orbit would probably be more useful than one in a polar orbit for:

a. relaying television signals across the Atlantic.
b. studying global weather patterns.
c. mapping the earth's major population centers.
d. locating ships lost at sea.

Name ______________________ Class ______________ Date ______________

MODERN EARTH SCIENCE

Section 3.1

Finding Locations on the Earth

Choose the one best response. Write the letter of that choice in the space provided.

_____ **1.** A plane leaves Greenwich, England, flying in a westerly direction. After flying three fourths of the way around the earth, the plane's longitude would be:

a. 90° E. **b.** 90° W. **c.** 270° E. **d.** 270° W.

_____ **2.** The circumference of the earth is approximately:

a. 400 km. **b.** 4,000 km. **c.** 40,000 km. **d.** 400,000 km.

_____ **3.** The earth's axis of rotation intersects the earth's surface at the:

a. magnetic poles. **b.** geographic poles.
c. prime meridian. **d.** equator.

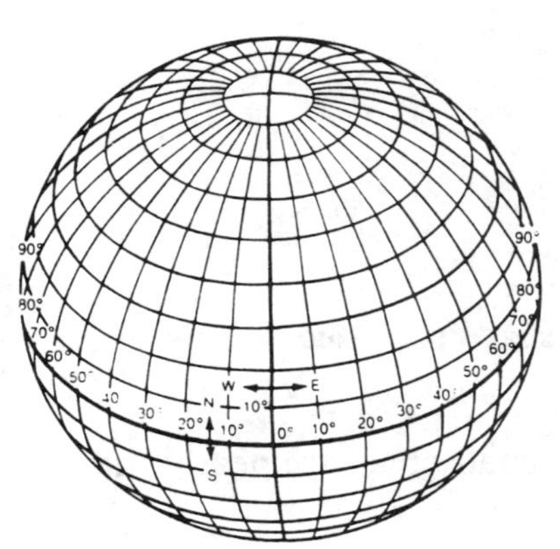

_____ **4.** The vertical lines in this diagram represent:

a. relief. **b.** parallels.
c. seconds. **d.** meridians.

_____ **5.** The latitude of the equator is:

a. 0°. **b.** 90°.
c. 180°. **d.** 360°.

Complete each statement by writing the correct term or phrase in the space provided.

6. Magnetic declination is the angle between the direction in which the compass needle points and the direction of the ______________________ .

7. An imaginary line that divides the earth into equal halves is called a ______________________ .

8. The line of longitude selected to be 0° is called the ______________________ .

9. True north is in the direction of the geographic ______________________ .

10. A degree of latitude consists of 60 equal parts called ______________________ .

Name ______________________ Class ______________ Date ______________

MODERN EARTH SCIENCE

Section 3.2

Mapping the Earth's Surface

Choose the one best response. Write the letter of that choice in the space provided.

_____ **1.** On a Mercator projection, the greatest distortion is produced:

a. at the equator. **b.** at the poles.
c. along the prime meridian. **d.** midway between the poles.

_____ **2.** Which type of map is produced by fitting together a series of map projections?

a. Mercator **b.** conic **c.** polyconic **d.** gnomonic

_____ **3.** Great circles are shown as straight lines on which type of map projection?

a. gnomonic **b.** Mercator **c.** polyconic **d.** conic

_____ **4.** Which of the following is true of a Mercator projection?

a. Coastline shapes are accurate.
b. Spacing between parallels is equal.
c. Compass directions are curved.
d. Equatorial regions are greatly distorted.

Read each statement below. If the statement is true, write *T* in the space provided. If the statement is false, write *F* in the space provided.

_____ **5.** On a polyconic projection, the relative size and shape of small areas are nearly the same as those on a globe.

_____ **6.** The globe is a type of map projection.

_____ **7.** The ratio of 1:10,000 can also be expressed as 1 cm equals 1 km.

_____ **8.** Directions on a map should be determined in relation to the meridians and parallels.

_____ **9.** According to this map, the distance between Osaka and Tokyo is approximately 500 km.

_____ **10.** Osaka is a capital city.

Name ______________________ Class ______________ Date ______________

MODERN EARTH SCIENCE

Section 3.3

Topographic Maps

Read each statement below. If the statement is true, write *T* in the space provided. If the statement is false, write *F* in the space provided.

_____ **1.** On topographic maps, contour lines connect points having the same elevation.

_____ **2.** Topographic maps show only natural features of an area.

_____ **3.** Eastern and western boundaries of U.S.G.S. maps are indicated by meridians of longitude.

_____ **4.** Index contours are contour lines that are printed bolder in order to make map reading easier.

_____ **5.** A 15' topographic map shows more detail than a 7.5' map of the same area.

Choose the one best response. Write the letter of that choice in the space provided.

_____ **6.** On a topographic map, rapid increases in elevation are represented by:

a. depression contours
b. V-shaped contour lines.
c. contour lines forming closed loops.
d. closely spaced contour lines.

_____ **7.** The elevation difference between two contour lines on a map is called the:

a. map scale. **b.** contour interval.
c. map projection. **d.** index contour.

_____ **8.** Contour intervals are most likely to be smallest on maps of:

a. hilly areas. **b.** flat areas.
c. high-altitude areas. **d.** low-altitude areas.

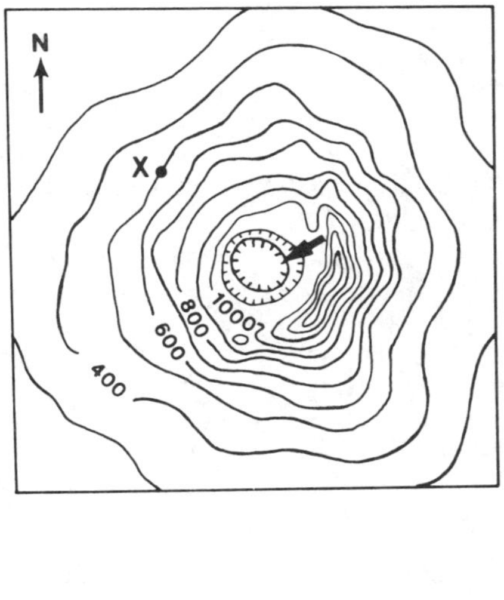

_____ **9.** At what elevation is the point labeled **X** on the map?

a. 400 m **b.** 450 m
c. 500 m **d.** 600 m

_____ **10.** Which feature is indicated by the arrow?

a. peak **b.** valley
c. stream **d.** depression

Name ________________ Class ________ Date ________

MODERN EARTH SCIENCE

Section 4.1

Continental Drift

Read each statement below. If the statement is true, write *T* in the space provided. If the statement is false, write *F* in the space provided.

_____ **1.** The magnetic orientation of a rock is determined when the rock solidifies.

_____ **2.** The theory of plate tectonics explains how some coastal mountain ranges may have been formed.

_____ **3.** All Mid-Atlantic Ridge rock samples examined have proven to be older than the oldest-known continental rocks.

_____ **4.** Magma is the molten rock that wells up through fissures in the earth's crust.

_____ **5.** The magnetic orientation of molten rock from a mid-ocean ridge will always reflect a reversed polarity of the earth's magnetic field at that time.

_____ **6.** Movement of the earth's crust away from an oceanic ridge is called seafloor spreading.

Choose the one best response. Write the letter of that choice in the space provided.

_____ **7.** Which scientist first proposed that the continents were once joined in a single landmass called Pangaea?

a. Hess **b.** Suess **c.** Dietz **d.** Wegener

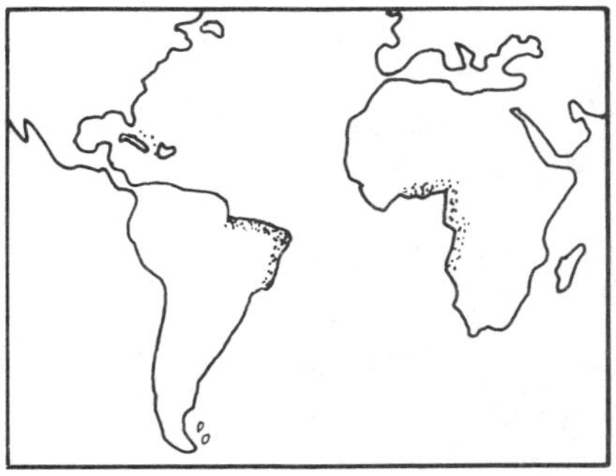

_____ **8.** Which theory is most directly supported by the discovery of rocks similar in age and type in the areas shaded on the map?

a. relativity **b.** suspect terranes
c. plate tectonics **d.** seafloor spreading

_____ **9.** When examining rocks from both sides of the Mid-Atlantic Ridge, scientists found evidence for the phenomenon of:

a. suspect terranes. **b.** magnetic reversal.
c. volcano formation. **d.** convection currents.

_____ **10.** What evidence found in tropical regions of southern Africa and South America most strongly supports the theory that the continents were once joined?

a. coal deposits **b.** mountain chains
c. glacial debris **d.** land bridges

Name ______________________ Class ______________ Date ______________

MODERN EARTH SCIENCE

Section 4.2

The Theory of Plate Tectonics

Choose the one best response. Write the letter of that choice in the space provided.

_____ **1.** Seafloor spreading occurs at which of the following plate boundaries?

a. divergent **b.** transform fault
c. convergent **d.** subduction

_____ **2.** The Palo Alto Hills are probably an example of:

a. a landmass bridge. **b.** an island arc.
c. a glacial landform. **d.** a suspect terrane.

_____ **3.** Which of the following may result from the collision of one plate with another?

a. a convergent boundary **b.** a divergent boundary
c. a rift valley **d.** a transform fault boundary

_____ **4.** Where in the earth do convection currents occur?

a. in the lithosphere **b.** along a subduction zone
c. in the asthenosphere **d.** along a rift valley

Complete each statement by writing the correct term or phrase in the space provided.

5. A fault formed at the point where two plates slide past each other is called a

______________________ .

6. Many scientists believe that the movement of lithospheric plates is caused by

______________________ .

7. The term for the type of plate boundary shown at the point labeled **X** in this diagram is

______________________ .

8. The layer labeled **Y** is the part of the lithosphere known as

the ______________________ .

9. The layer of plastic rock beneath the lithosphere is called the ______________________ .

10. The theory that proposes a possible explanation of why and how continents move is called the theory of ______________________ .

Name ____________ Class ____________ Date ____________

Section 5.1

How the Crust Is Deformed

Read each statement below. If the statement is true, write *T* in the space provided. If the statement is false, write *F* in the space provided.

_____ **1.** When a glacier retreats from an area, the part of the crust that was formerly ice-covered actually sinks deeper into the mantle.

_____ **2.** Stress in the crust can be caused by isostatic adjustments.

_____ **3.** Forces that cause deformation of the crust are usually the result of a change in the volume of the mantle.

_____ **4.** A very thick deposit of material on the ocean floor is called a hot spot.

_____ **5.** Compression causes crustal rocks to be squeezed together.

_____ **6.** Isostatic adjustments are constantly occurring in areas of the earth's crust with mountain ranges.

Choose the one best response. Write the letter of that choice in the space provided.

_____ **7.** Up-and-down motions of the crust are called:

a. thrust faulting. **b.** strain movement.
c. isostatic adjustments. **d.** compressional stress.

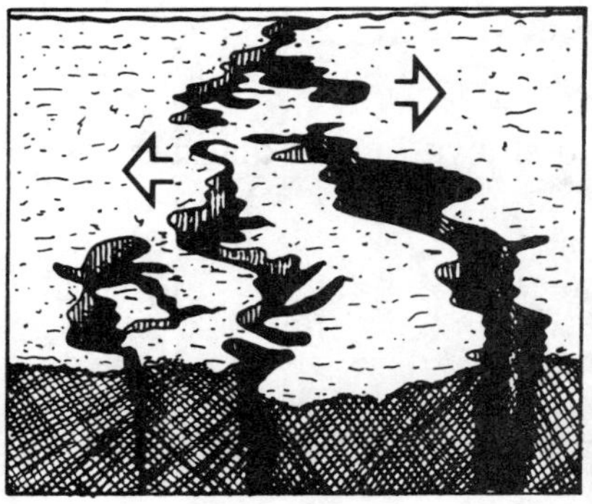

_____ **8.** What type of force has acted on the rocks in this diagram?

a. shearing **b.** tension
c. compression **d.** strain

_____ **9.** A change in the shape or volume of crustal rock due to stress is called:

a. shearing. **b.** isostasy. **c.** compression. **d.** strain.

_____ **10.** As rivers flow into the ocean and deposit thick layers of sediment on the ocean floor, what will the crust under the sediment do?

a. rise **b.** fracture **c.** sink **d.** erupt

Section 5.2

The Results of Stress

Read each statement below. If the statement is true, write *T* in the space provided. If the statement is false, write *F* in the space provided.

______ **1.** Normal faults form in regions where the crust is diverging.

______ **2.** Along a strike-slip fault, the rock on either side of the fault plane moves vertically.

______ **3.** A reverse fault occurs when compression causes a hanging wall to move up relative to a footwall.

______ **4.** A thrust fault has no hanging wall.

______ **5.** A thrust fault is a type of normal fault.

Complete each statement by writing the correct term or phrase in the space provided.

6. The actual surface of a break in the crustal rocks at a fault is called a fault ______________________ .

7. The crustal rocks below the broken surface at a normal fault make up the ______________________ .

8. The term for the type of fold shown in this diagram is ______________________ .

9. When rocks respond to stress by becoming permanently deformed but not breaking, the result is the process of ______________________ .

10. Downcurved folds in layered rock are called ______________________ .

Name ______________________ Class ______________ Date ____________

Section 5.3

Mountain Formation

Read each statement below. If the statement is true, write *T* in the space provided. If the statement is false, write *F* in the space provided.

______ **1.** The Himalayas are dome mountains.

______ **2.** The circum-Pacific belt is primarily an area of converging plate boundaries.

______ **3.** The suspect terrane theory relates to small pieces of the crust that have been scraped from subducting oceanic crust.

______ **4.** The Appalachian Mountains are fault-block mountains.

______ **5.** Most plateaus are formed when thick, horizontal layers of rock are slowly uplifted.

______ **6.** According to the theory of plate tectonics, India was once a separate continent riding on the Indian plate.

Choose the one best response. Write the letter of that choice in the space provided.

______ **7.** What type of mountains are commonly located where continents have collided?

a. folded **b.** volcanic **c.** fault-block **d.** dome

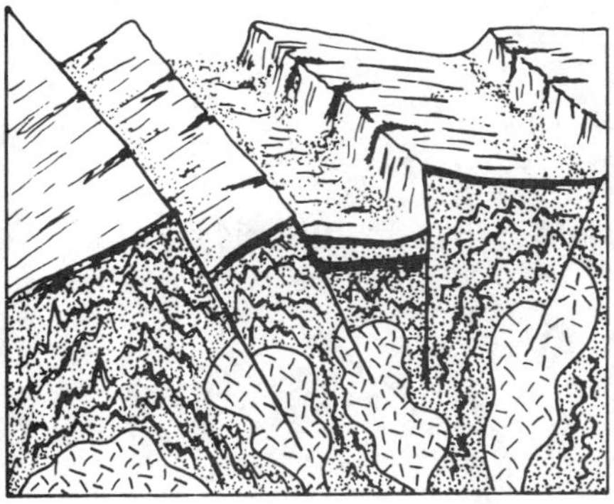

______ **8.** What type of mountains are shown in this diagram?

a. folded
b. volcanic
c. dome
d. fault-block

______ **9.** What are the largest groups of mountains on earth called?

a. ridges **b.** ranges **c.** belts **d.** systems

______ **10.** A collision between continental and oceanic crust formed the:

a. Mid-Atlantic Ridge. **b.** Cascade Mountains.
c. Himalayan Mountains. **d.** Adirondack Mountains.

Name ________________ Class ________________ Date ________________

Section 6.1

Earthquakes and Plate Tectonics

Choose the one best response. Write the letter of that choice in the space provided.

______ **1.** When friction prevents the rocks on each side of a fault from moving past each other, the fault is said to be:

a. fractured. **b.** subducting. **c.** locked. **d.** elastic.

______ **2.** The place where slippage first occurs is called an earthquake's:

a. focus. **b.** epicenter. **c.** magnitude. **d.** intensity.

______ **3.** In which major earthquake zone are earthquakes produced mainly by plates moving away from each other?

a. the Pacific Ring of Fire
b. the Mid-Atlantic Ridge
c. the San Andreas Fault area
d. the Eurasian-Melanesian mountain belt

______ **4.** Most earthquakes occur along or near the edges of the:

a. North American plate. **b.** earth's oceans and lakes.
c. Eurasian plate. **d.** earth's lithospheric plates.

______ **5.** Which type of earthquake usually occurs farther inland than other earthquakes?

a. deep-focus **b.** shallow-focus
c. intermediate-focus **d.** microquakes

Complete each statement by writing the correct term or phrase in the space provided.

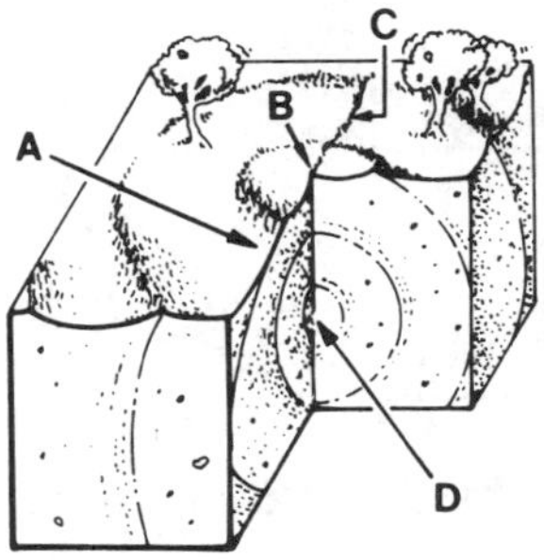

A
B
C
D

6. The point that indicates the epicenter of the earthquake in the diagram is ________________ .

7. The point that indicates the focus of the earthquake is ________________ .

8. A group of interconnected faults is called a ________________ .

9. Small tremors following an earthquake are called ________________ .

10. When fracturing and slipping into new positions, rocks along a fault release energy in the form of vibrations called ________________ .

Name ______________________ Class ______________ Date ______________

Section 6.2

Recording Earthquakes

Read each statement below. If the statement is true, write *T* in the space provided. If the statement is false, write *F* in the space provided.

_____ **1.** Microquakes are usually strong enough to be felt by people.

_____ **2.** A seismograph records energy released by an earthquake.

_____ **3.** Scientists analyze the difference between the arrival times of P and S waves to determine an earthquake's epicenter.

_____ **4.** P waves moving through the earth can travel through solids, liquids, and gases.

_____ **5.** An earthquake with the highest magnitude is given a rating of 31.7 on the Richter scale.

Complete each statement by writing the correct term or phrase in the space provided.

6. The scale that is used to measure the intensity of an earthquake is called the

______________________ .

7. An earthquake with a magnitude of 2 is classified as a ______________________ .

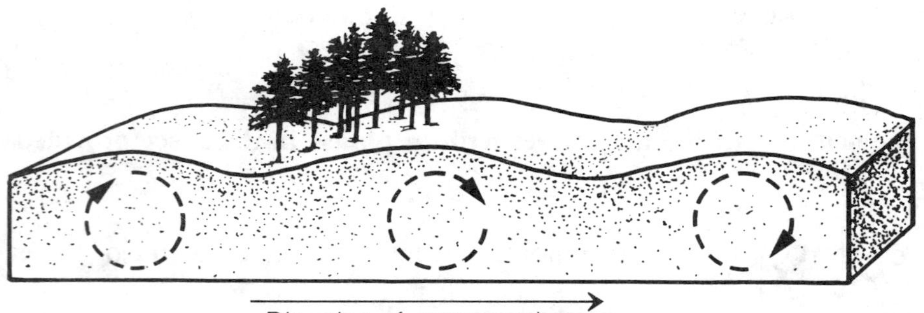

Direction of wave motion

8. The seismic waves shown in this diagram are called ______________________ .

9. Seismic waves that cause rock particles to move at right angles to the direction in which the waves are traveling are called ______________________ .

10. The amount of energy released by an earthquake is also known as its ______________________ .

Name ______________________ Class ______________ Date ______________

Section 6.3

Earthquake Damage

Read each statement below. If the statement is true, write *T* in the space provided. If the statement is false, write *F* in the space provided.

_____ **1.** The epicenter of an earthquake that causes a tsunami is located on the ocean floor.

_____ **2.** Movement of the ground is the direct cause of most injuries related to earthquakes.

_____ **3.** Earthquake predictions based on records of past earthquakes in the area can be off by years.

_____ **4.** Following simple safety rules during an earthquake can help prevent death and injury.

_____ **5.** Duration does not affect the intensity of an earthquake.

Choose the one best response. Write the letter of that choice in the space provided.

_____ **6.** In general, where is the safest place for an individual to park when in a car during an earthquake?

a. close to water
b. in a tunnel
c. in the open
d. under a bridge

A
Ocean floor
B
C
D

_____ **7.** Which point in this diagram indicates a likely source of a tsunami?

a. A **b.** B **c.** C **d.** D

_____ **8.** An earthquake is often preceded by a decrease in:

a. the amount of gas seepage from fractured rock.
b. the intensity of local L waves.
c. the degree to which the ground tilts.
d. the speed of local P waves.

_____ **9.** Zones of immobile rock along faults are called:

a. epicenters.
b. seismic gaps.
c. ridges.
d. subduction zones.

_____ **10.** The Seismic Sea Wave Warning System (SSWWS) issues warnings of:

a. tsunamis. **b.** earthquakes. **c.** aftershocks. **d.** landslides.

Section 7.1

Volcanoes and Plate Tectonics

Read each statement below. If the statement is true, write *T* in the space provided. If the statement is false, write *F* in the space provided.

_____ **1.** Solid rock located deep in the mantle is called magma.

_____ **2.** Magma forms at subducted plate boundaries.

_____ **3.** A major zone of active volcanoes encircles the Atlantic Ocean.

Choose the one best response. Write the letter of that choice in the space provided.

_____ **4.** Knowledge of temperature and pressure conditions in the earth's mantle is based on the analysis of:

a. direct measurements. **b.** seismic wave behavior.
c. crustal magnetic properties. **d.** volcanic bulges.

_____ **5.** An opening in the earth's surface through which molten rock flows is called a:

a. vent. **b.** caldera. **c.** mantle. **d.** fault.

_____ **6.** Great pressure in the asthenosphere keeps the rock there mostly:

a. felsic. **b.** molten. **c.** solid. **d.** gaseous.

_____ **7.** An opening on the earth's surface through which molten rock flows and the material that builds up around it forms a:

a. subduction zone. **b.** trench.
c. convergent boundary. **d.** volcano.

Hawaiian Islands
X
Oceanic crust

_____ **8.** The area labeled **X** in the diagram is located over:

a. a caldera. **b.** a hot spot.
c. a mid-ocean ridge. **d.** an island arc.

_____ **9.** A string of volcanoes that forms along a trench is called:

a. an island arc. **b.** a fissure.
c. a mid-ocean ridge. **d.** a subducted plate.

_____ **10.** Iceland was formed over a:

a. mid-ocean ridge. **b.** subduction zone.
c. convergent boundary. **d.** caldera.

Name ______________________ Class ______________ Date ______________

Section 7.2

Volcanic Eruptions

Choose the one best response. Write the letter of that choice in the space provided.

_____ **1.** Cinder cones are generally formed when volcanoes eject:

a. thin lava. **b.** solid fragments. **c.** pillow lava. **d.** gases.

_____ **2.** What type of lava usually flows out of oceanic volcanoes?

a. felsic **b.** aa **c.** mafic **d.** pahoehoe

_____ **3.** Felsic lava is usually identified by its color and:

a. silica content. **b.** rounded shapes. **c.** ropey texture. **d.** sulfur content.

_____ **4.** Which of the following generally results from a quiet lava eruption?

a. cinder cone **b.** surface bulge **c.** shield cone **d.** stratovolcano

_____ **5.** Before a volcanic eruption, seismic activity seems to:

a. increase in frequency and decrease in intensity.
b. decrease in both frequency and intensity.
c. decrease in frequency and increase in intensity.
d. increase in both frequency and intensity.

_____ **6.** What causes bulging in the surface of volcanoes?

a. buildup of volcanic blocks **b.** upward movement of magma **c.** changes in volcanic gases **d.** tephra eruptions

Complete each statement by writing the correct term or phrase in the space provided.

VOLCANIC ROCK FRAGMENTS

Size	Type
less than 0.25 mm	dust
0.25 – 2 mm	
2 – 64 mm	*
more than 64 mm	bombs

7. The fragment type indicated by the asterisk in the table is ______________________ .

8. Volcanic rock particles small enough to be carried by wind are dust and ______________________ .

9. Round or spindle-shaped tephra is called ______________________ .

10. The basin-shaped depression left by a volcanic explosion is called a ______________________ .

Name ______________________ Class ______________ Date ______________

MODERN EARTH SCIENCE

Section 7.3

Extraterrestrial Volcanism

Read each statement below. If the statement is true, write *T* in the space provided. If the statement is false, write *F* in the space provided.

_____ **1.** Plate tectonics was probably the primary cause of volcanism on the moon.

_____ **2.** Meteorite bombardment of the moon may have generated the heat necessary to form magma.

_____ **3.** The moon Io has only one tenth the volcanic activity of the earth.

Choose the one best response. Write the letter of that choice in the space provided.

_____ **4.** What type of volcanic formation is Olympus Mons?

a. shield cone
b. cinder cone
c. stratovolcano
d. oceanic volcano

_____ **5.** What type of lava flow has been found on the earth's moon?

a. pahoehoe
b. pillow
c. felsic
d. basaltic

_____ **6.** Which of the following planetary bodies has far more volcanic activity than the earth?

a. Mars
b. the moon
c. Io
d. Jupiter

_____ **7.** Scientists believe that the material erupting from Io's volcanoes is primarily:

a. felsic material.
b. sulfur and sulfur dioxide.
c. mafic material.
d. iron and carbon dioxide.

_____ **8.** As a result of volcanic activity, Io's surface is colored:

a. jet black.
b. ashy gray.
c. bright yellow-red.
d. dull yellow-brown.

_____ **9.** The heat that causes the melting of Io's interior is probably generated by:

a. friction from tidal movement.
b. solar radiation.
c. constant meteorite bombardment.
d. tectonic activity.

_____ **10.** Where is the volcano Olympus Mons located?

a. Io
b. the moon
c. Saturn
d. Mars

Name ______________________ Class ______________ Date ______________

Section 8.1

Matter

Read each statement below. If the statement is true, write *T* in the space provided. If the statement is false, write *F* in the space provided.

______ **1.** Physical properties describe how a substance interacts with other substances to produce different kinds of matter.

______ **2.** Elements cannot be broken down into simpler forms by ordinary chemical means.

______ **3.** The particles of a solid are more tightly packed than those of a gas.

______ **4.** Isotopes are atoms of the same element that differ in atomic number.

______ **5.** The particles in a liquid move faster when heated.

Choose the one best response. Write the letter of that choice in the space provided.

______ **6.** Which of the following has a negative charge?

a. neutron **b.** nucleus **c.** proton **d.** electron

______ **7.** Atoms of an element with an atomic number of 8 must have:

a. 8 neutrons. **b.** 8 protons. **c.** 16 neutrons. **d.** 16 protons.

______ **8.** Different elements are made up of different kinds of:

a. electrons. **b.** protons. **c.** molecules. **d.** atoms.

63.55 Cu Copper 29 (2, 8, 18, 1)	65.39 Zn Zinc 30 (2, 8, 18, 2)	69.72 Ga Gallium 31 (2, 8, 18, 3)
107.9 Ag Silver 47 (2, 8, 18, 18, 1)	112.4 Cd Cadmium 48 (2, 8, 18, 18, 2)	114.8 In Indium 49 (2, 8, 18, 18, 3)
197.0 Au Gold 79 (2, 8, 18, 32, 18, 1)	200.6 Hg Mercury 80 (2, 8, 18, 32, 18, 2)	204.4 Tl Thallium 81 (2, 8, 18, 32, 18, 3)

______ **9.** According to this table, how many electrons does the element silver (Ag) have?

a. 47 **b.** 61
c. 108 **d.** 148

______ **10.** How many neutrons could an atom of the element zinc (Zn) have?

a. 30 **b.** 36
c. 65 **d.** 95

Name ______________________ Class ______________ Date ______________

MODERN EARTH SCIENCE

Section 8.2

Combinations of Atoms

Complete each statement by writing the correct term or phrase in the space provided.

1. The smallest complete unit of a compound is a ______________________ .
2. The written notation used to represent the elements a compound contains and the relative number of atoms of each element is called a ______________________ .
3. Atoms in all compounds are held together by forces called ______________________ .
4. A mixture in which one substance is uniformly distributed in another substance is called a ______________________ .
5. Molecules made up of two atoms are called ______________________ .

Choose the one best response. Write the letter of that choice in the space provided.

_____ **6.** Which of the following is an alloy?

a. rubber **b.** bronze **c.** wood **d.** iron

_____ **7.** Which of the following is shared between atoms in a covalent bond?

a. ions **b.** neutrons **c.** molecules **d.** electrons

_____ **8.** What is the maximum number of electrons an element with two energy levels can have?

a. 2 **b.** 4 **c.** 8 **d.** 10

_____ **9.** How many atoms of potassium (K) are represented by the formula K_2SO_4?

a. 1 **b.** 2 **c.** 6 **d.** 7

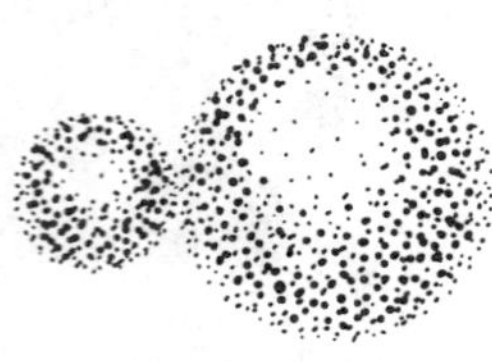

_____ **10.** According to this diagram, which of the following most accurately describes the formation of sodium chloride (NaCl)?

a. Chlorine gains an electron.
b. Sodium gains a proton.
c. Chlorine loses an electron.
d. Sodium loses a proton.

Name ______________________ Class ______________ Date ______________

MODERN EARTH SCIENCE

Section 9.1

What Is a Mineral?

Read each statement below. If the statement is true, write *T* in the space provided. If the statement is false, write *F* in the space provided.

_____ **1.** Silicate minerals make up more than 90 percent of the earth's crust.

_____ **2.** The hardness of quartz is due primarily to bonds between oxygen and aluminum atoms.

_____ **3.** Plagioclase feldspar may contain sodium, calcium, or both.

_____ **4.** Minerals are organic solids that are formed in the earth.

_____ **5.** Sulfates and sulfides are types of nonsilicate minerals.

Choose the one best response. Write the letter of that choice in the space provided.

_____ **6.** Minerals made up of single chains of Si-O tetrahedra are called:

a. pyroxenes. **b.** amphiboles. **c.** crystals. **d.** sheets.

_____ **7.** Which of the following is an example of a mineral?

a. coal **b.** concrete **c.** steel **d.** quartz

_____ **8.** The minerals gold and copper are examples of:

a. silicate minerals.
b. organic compounds.
c. native elements.
d. radioactive substances.

_____ **9.** How many oxygen atoms are in a silicon-oxygen tetrahedron?

a. 2 **b.** 3 **c.** 4 **d.** 5

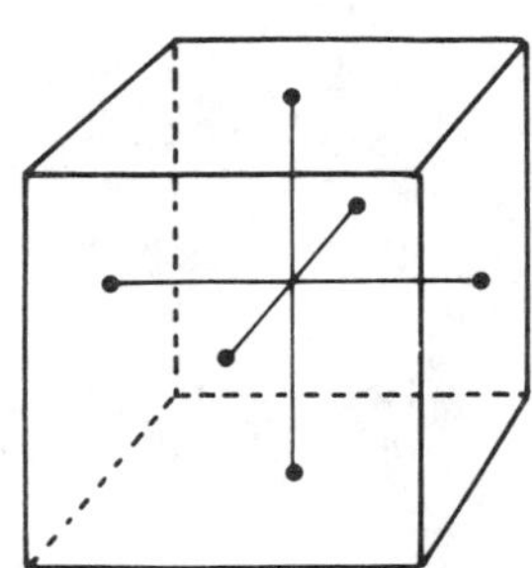

_____ **10.** The structure presented in the diagram is an example of a:

a. tetragonal crystal.
b. hexagonal crystal.
c. triclinic crystal.
d. cubic crystal.

Section 9.2

Identifying Minerals

Read each statement below. If the statement is true, write *T* in the space provided. If the statement is false, write *F* in the space provided.

_____ **1.** Light rays are refracted or bent as they pass from one substance to a different substance.

_____ **2.** The streak test is a test of mineral density.

_____ **3.** Fluorescent minerals glow while being subjected to ultraviolet light.

_____ **4.** Cleavage is important in identifying some minerals.

_____ **5.** All minerals have the same basic crystal shape.

Choose the one best response. Write the letter of that choice in the space provided.

_____ **6.** The heft or relative weight of a mineral sample is directly related to the mineral's:

a. luster. **b.** cleavage. **c.** density. **d.** hardness.

_____ **7.** Needles of early compasses were often constructed of:

a. lodestone. **b.** calcite. **c.** talc. **d.** uranium.

_____ **8.** Which of the following is an ore containing unstable nuclei?

a. gypsum **b.** feldspar **c.** magnetite **d.** pitchblende

Mineral	Hardness
feldspar	6
quartz	7
topaz	8
corundum	9

_____ **9.** According to the table, what is the approximate hardness of a mineral that scratches quartz and can be scratched by topaz?

a. 6.5 **b.** 7.5
c. 8.5 **d.** 9.5

_____ **10.** Diamond has which of the following types of luster?

a. metallic **b.** brilliant **c.** glassy **d.** pearly

Name ______________________ Class ______________ Date ______________

Section 10.1

Rocks and the Rock Cycle

Read each statement below. If the statement is true, write *T* in the space provided. If the statement is false, write *F* in the space provided.

______ **1.** The three types of rock are igneous, sedimentary, and extrusive.

______ **2.** Any one of the three types of rock can be changed into any other type.

______ **3.** Rocks are classified according to how they were formed.

______ **4.** Sedimentary rock is the parent material for all rocks.

______ **5.** High temperatures can change sedimentary rock directly into magma.

Complete each statement by writing the correct term or phrase in the space provided.

6. Rocks are changed from one type to another in a series of changes called the rock

______________________ .

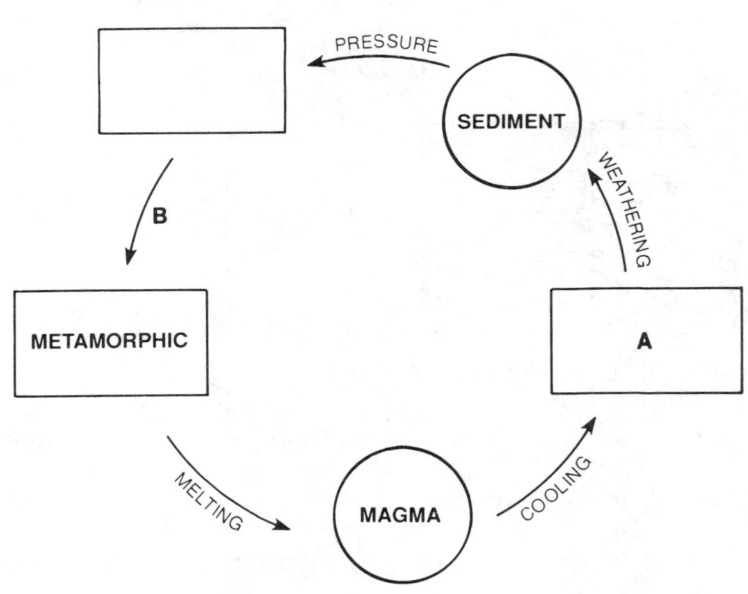

7. The type of rock represented by rectangle **A** in the diagram is ______________________ .

8. The arrow labeled **B** represents heat and ______________________ .

9. Rocks, minerals, and organic matter that have been broken into fragments are called

______________________ .

10. Magma that reaches the earth's surface is called ______________________ .

Name ______________________ Class ______________ Date ______________

Section 10.2

Igneous Rock

Read each statement below. If the statement is true, write *T* in the space provided. If the statement is false, write *F* in the space provided.

_____ **1.** Mafic rocks are light-colored rocks that are high in silica.

_____ **2.** Laccoliths often form small, dome-shaped mountains.

_____ **3.** A sill is an igneous structure that cuts across rock layers.

_____ **4.** Batholiths are metamorphic rock structures.

_____ **5.** Volcanic glass forms when magma cools very quickly.

Choose the one best response. Write the letter of that choice in the space provided.

_____ **6.** An igneous rock with a mixture of large and small grains is:

a. a porphyry. **b.** an intrusion. **c.** an extrusion. **d.** a breccia.

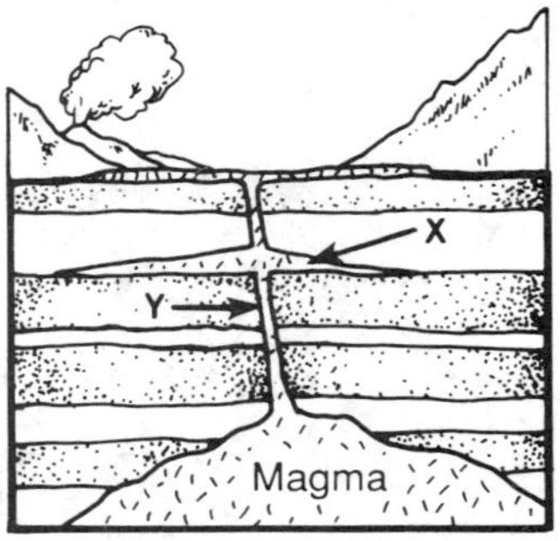

X

Y

Magma

_____ **7.** The structure labeled **X** in the diagram is a:

a. sill. **b.** dike. **c.** stock. **d.** batholith.

_____ **8.** The structure labeled **Y** in the diagram is a:

a. sill. **b.** dike. **c.** laccolith. **d.** batholith.

_____ **9.** Which of the following is one of the three families of igneous rocks?

a. clastic **b.** foliated **c.** gabbro **d.** diorite

_____ **10.** Felsic rocks are high in:

a. quartz. **b.** silica. **c.** biotite. **d.** calcite.

Section 10.3

Sedimentary Rock

Read each statement below. If the statement is true, write *T* in the space provided. If the statement is false, write *F* in the space provided.

_____ **1.** Some sedimentary rocks are formed from the remains of organisms.

_____ **2.** Halite is a clastic sedimentary rock.

_____ **3.** Fossils are most commonly preserved in sedimentary rock.

_____ **4.** Shale consists of clay-sized particles that are pressed into flat layers.

_____ **5.** Stratification occurs when the same particle type is deposited over a long period of time.

Complete each statement by writing the correct term or phrase in the space provided.

6. Sedimentary rock is formed by the processes of cementation and

______________________ .

7. Sedimentary rocks formed by minerals that are left behind when water evaporates are called ______________________ .

8. The crystal-filled rock in the diagram is a

______________________ .

9. Nodules that form when minerals precipitated from solutions build up around rock particles are called

______________________ .

10. Limestone can form either as an organic sedimentary rock or as a

______________________ .

Name ______________________ Class ______________ Date ______________

Section 10.4

Metamorphic Rock

Read each statement below. If the statement is true, write *T* in the space provided. If the statement is false, write *F* in the space provided.

_____ **1.** Most metamorphic rock forms near the earth's surface.

_____ **2.** Metamorphism can change the mineral composition of rock.

_____ **3.** Most metamorphic rock is formed by tectonic activity.

_____ **4.** Metamorphism results from heat, pressure, and chemical processes.

_____ **5.** Metamorphism can create a hard and durable rock from a softer sedimentary rock.

Complete each statement by writing the correct term or phrase in the space provided.

6. Quartzite is produced by the metamorphism of ______________________ .

7. The type of metamorphism that results from the heat of magma is called

______________________ .

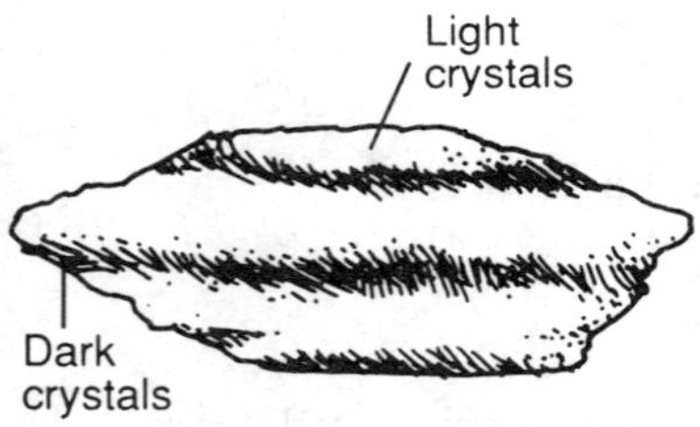

Light crystals

Dark crystals

8. The appearance of the metamorphic rock in the diagram indicates that it should be classified as

______________________ .

9. Slate is formed when great pressure acts on the sedimentary rock

______________________ .

10. Metamorphism that occurs over large areas is called ______________________ .

Name ______________________ Class ______________ Date ______________

Section 11.1

Mineral Resources

Read each statement below. If the statement is true, write *T* in the space provided. If the statement is false, write *F* in the space provided.

_____ **1.** A lode is a layer of minerals that has been deposited in a stream bed.

_____ **2.** The ocean floor is a largely untapped source of mineral resources.

_____ **3.** Minerals are renewable resources.

Choose the one best response. Write the letter of that choice in the space provided.

_____ **4.** Hematite is a source of:

a. iron. **b.** aluminum. **c.** zinc. **d.** diamonds.

_____ **5.** Which of the following is a native element?

a. chalcopyrite **b.** copper **c.** bauxite **d.** hydrocarbon

_____ **6.** Contact metamorphism is caused primarily by:

a. heat. **b.** rain.
c. radioactivity. **d.** wind.

_____ **7.** Which of the following is a source of glass?

a. bauxite **b.** sphalerite **c.** calcite **d.** quartz

_____ **8.** At which location in the diagram would a placer deposit most likely be found?

a. A **b.** B
c. C **d.** D

_____ **9.** Minerals that have shiny surfaces and are good conductors of electricity are:

a. ores. **b.** metals. **c.** lodes. **d.** gemstones.

_____ **10.** Recycling is a means of:

a. conserving minerals. **b.** mining minerals.
c. locating minerals. **d.** identifying minerals.

Name ______________________ Class ______________ Date ______________

MODERN EARTH SCIENCE

Section 11.2

Fossil Fuels

Read each statement below. If the statement is true, write *T* in the space provided. If the statement is false, write *F* in the space provided.

_____ **1.** Synthetic rubber is commonly made from bauxite.

_____ **2.** Coal is produced by the process of carbonization.

_____ **3.** Sulfur dioxide is often released into the atmosphere when coal is burned.

Choose the one best response. Write the letter of that choice in the space provided.

_____ **4.** The process of coal formation began in ancient:

a. magma bodies. **b.** oceans.
c. swamps. **d.** caves.

_____ **5.** Anthracite is a type of:

a. geothermal energy source. **b.** organic rock.
c. gemstone. **d.** microorganism.

_____ **6.** Which of the following is produced by the accumulation of microorganisms on the ocean floor?

a. natural gas **b.** gold deposits **c.** uranium **d.** coal

_____ **7.** The structure labeled **X** in the diagram is:

a. a fault. **b.** a cap rock.
c. a vein. **d.** a lode.

_____ **8.** Crude oil is an important source of:

a. geothermal energy. **b.** petrochemicals.
c. cinnabar. **d.** anthracite.

_____ **9.** According to scientists, what percentage of the petroleum in the United States has already been discovered?

a. 5% **b.** 25% **c.** 50% **d.** 75%

_____ **10.** Strip mining often causes harm to the environment by exposing:

a. acidic rocks. **b.** radioactive elements.
c. natural gas. **d.** sulfur compounds.

Name ______________________ Class ______________ Date ______________

Section 11.3

Nuclear Energy

Read each statement below. If the statement is true, write *T* in the space provided. If the statement is false, write *F* in the space provided.

_____ **1.** Nuclear fuel produces radiation that is harmful to living cells.

_____ **2.** Uranium-238 is the radioactive isotope used for nuclear fission.

_____ **3.** Nuclear fusion is the combining of atomic nuclei.

Choose the one best response. Write the letter of that choice in the space provided.

_____ **4.** Which of the following is an important product of fission reactions?

a. sound energy
b. electrical energy
c. chemical energy
d. heat energy

_____ **5.** In nuclear reactors, electricity is produced when the turbines of generators are turned by:

a. neutrons. **b.** steam. **c.** protons. **d.** fuel pellets.

_____ **6.** A common method used to dispose of nuclear waste is to store it in:

a. cement vaults.
b. mountain caves.
c. salt mines.
d. coal mines.

_____ **7.** Which of the following reactions is represented by this diagram?

H + H → H H → He

a. decomposition
b. carbonation
c. nuclear fission
d. nuclear fusion

_____ **8.** Which of the following is a potential source of fuel for nuclear fusion reactions?

a. water **b.** coal **c.** uranium **d.** petroleum

_____ **9.** One advantage of nuclear fusion over nuclear fission as an energy source is that nuclear fusion:

a. is easily harnessed.
b. produces less harmful waste.
c. requires no fuel.
d. does not generate heat.

_____ **10.** Fission reactors are often cooled by circulating:

a. air. **b.** oxygen. **c.** water. **d.** oil.

Name ______________________ Class ______________ Date ______________

Section 11.4

Alternative Energy Sources

Choose the one best response. Write the letter of that choice in the space provided.

_____ **1.** Hydroelectric energy is energy produced by:

a. hot geysers. **b.** burning coal.
c. fission reactions. **d.** running water.

_____ **2.** What percentage of the electricity used in the United States comes from hydroelectric plants?

a. 4% **b.** 11% **c.** 27% **d.** 53%

_____ **3.** In power plants, the force of steam is used to spin turbines primarily in order to:

a. generate electricity. **b.** cool the reactor.
c. generate heat. **d.** drive machinery.

_____ **4.** A disadvantage of wind-driven generators is that:

a. fuel is expensive.
b. wind energy is inefficient.
c. the wind does not always blow.
d. generators pollute the atmosphere.

_____ **5.** Water heated as it flows through hot rocks is a source of:

a. geothermal energy. **b.** fossil fuels.
c. solar energy. **d.** hydroelectric power.

_____ **6.** Power plants using geothermal energy are used extensively in:

a. England. **b.** Iceland. **c.** Mexico. **d.** Canada.

Read each question and answer it in the space provided.

7. What type of energy is the building in the diagram designed to take advantage of? ______________________

8. What is the name for a rooftop box designed to capture the energy in sunlight? ______________________

9. Which type of energy makes use of the energy found in magma? ______________________

10. What are the two types of solar energy systems? ______________________

Section 12.1

Weathering Processes

Read each statement below. If the statement is true, write *T* in the space provided. If the statement is false, write *F* in the space provided.

_____ **1.** Rainwater is naturally acidic.

_____ **2.** Abrasion is a chemical process.

_____ **3.** Plants can act as both mechanical and chemical weathering agents.

_____ **4.** Chemical weathering breaks down a rock without changing the composition of the rock.

_____ **5.** Joints usually develop in granite when the pressure on it increases.

Choose the one best response. Write the letter of that choice in the space provided.

_____ **6.** What type of weathering most likely caused the formations in the diagram?

a. exfoliation **b.** abrasion
c. carbonation **d.** oxidation

_____ **7.** Weathering in which thin sheets of rock flake off the surface is called:

a. carbonation. **b.** decomposition.
c. oxidation. **d.** exfoliation.

_____ **8.** The rounding of a rock due to collision with other rocks is called:

a. abrasion. **b.** gullying.
c. exfoliation. **d.** creep.

_____ **9.** Feldspar can be changed by hydrolysis into:

a. limestone. **b.** iron oxide.
c. kaolin. **d.** carbonic acid.

_____ **10.** Lichens and mosses erode rocks by producing:

a. limestone. **b.** oxides. **c.** bases. **d.** acids.

Name ______________________ Class ______________ Date ______________

MODERN EARTH SCIENCE

Section 12.2

Rates of Weathering

Choose the one best response. Write the letter of that choice in the space provided.

_____ **1.** The rate at which conglomerates weather is most dependent upon the rock's:

a. cementing material. **b.** age.
c. grain size. **d.** texture.

_____ **2.** Which of the following best accounts for the damage to Cleopatra's Needle after being moved to New York City?

a. carbonation **b.** dry climate **c.** pollution **d.** cold climate

_____ **3.** Weathering is generally slow in climates with extended periods of:

a. acid rain. **b.** cold. **c.** high winds. **d.** humidity.

A B C D

_____ **4.** At which point in the diagram will weathering occur most rapidly?

a. A **b.** B
c. C **d.** D

Complete each statement by writing the correct term or phrase in the space provided.

5. The four main factors that determine the rate of weathering are climate, composition, exposure, and ______________________ .

6. Fractures and joints allow weathering to occur more rapidly by increasing a rock's ______________________ .

7. Water that penetrates rock and then freezes causes weathering by the process of ______________________ .

8. When a rock is split into eight smaller blocks, its surface area increases by a factor of ______________________ .

9. Rocks containing calcite are most easily weathered by the chemical process of ______________________ .

10. Carbonation is responsible for converting calcite to ______________________ .

Name ______________________ Class ______________ Date ______________

Section 12.3

Weathering and Soil

Read each statement below. If the statement is true, write *T* in the space provided. If the statement is false, write *F* in the space provided.

_____ **1.** Desert soil consists mostly of regolith.

_____ **2.** Transported soils have well-defined horizons.

_____ **3.** Pedocal soils are less acidic than pedalfer soils.

_____ **4.** Arctic soils are produced mainly by mechanical weathering.

_____ **5.** Weathered granite forms sandy soils.

Complete each statement by writing the correct term or phrase in the space provided.

6. Thick soil that forms in tropical climates is called ______________________ .

7. A dark, organic material in the soil produced by the decaying remains of plants and animals is called ______________________ .

8. The layers of a soil profile are called ______________________ .

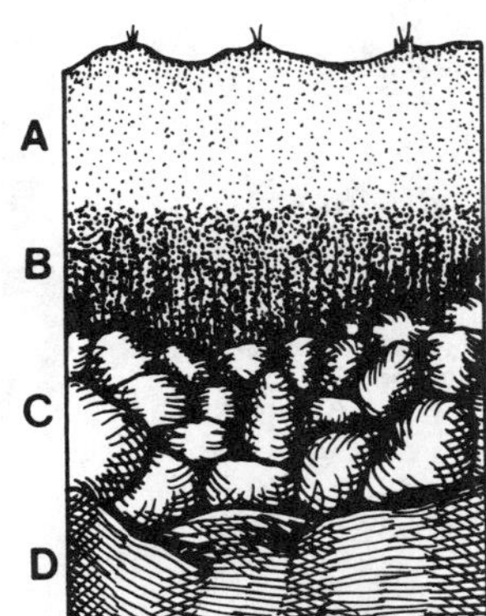

9. In this diagram, organic matter is most concentrated in the layer labeled ______________________ .

10. Bedrock is represented by the layer labeled ______________________ .

Section 12.4

Erosion

Read each statement below. If the statement is true, write *T* in the space provided. If the statement is false, write *F* in the space provided.

_____ **1.** Gullying often occurs when a slope is plowed for farming.

_____ **2.** Stripcropping generally accelerates erosion.

_____ **3.** Chemical and mechanical are two types of erosion.

_____ **4.** Rockfalls and landslides are examples of mass movements.

_____ **5.** Gravity is an important agent of erosion.

Choose the one best response. Write the letter of that choice in the space provided.

_____ **6.** A pile of rocks that accumulates at the base of a slope is called:

a. humus. **b.** slump. **c.** regolith. **d.** talus.

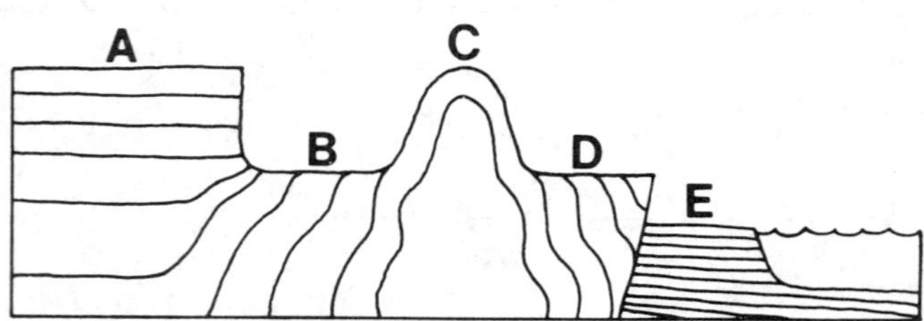

_____ **7.** In this diagram, which area represents a plateau?

a. A **b.** B **c.** D **d.** E

_____ **8.** The feature labeled **C** represents a:

a. butte. **b.** monadnock. **c.** mesa. **d.** peneplain.

_____ **9.** A mass movement in which the upper soil layers thaw and flow is called:

a. slump. **b.** solifluction. **c.** creep. **d.** mudflow.

_____ **10.** Small, tablelike areas that once formed a broad, flat landform are called:

a. mesas. **b.** plateaus. **c.** peneplains. **d.** terraces.

Name ______________________ Class ______________ Date ______________

Section 13.1

The Water Cycle

Read each statement below. If the statement is true, write *T* in the space provided. If the statement is false, write *F* in the space provided.

_____ **1.** The hydrologic cycle is also called the water cycle.

_____ **2.** Most water evaporating from the earth's surface evaporates from rivers and lakes.

_____ **3.** When water vapor rises in the atmosphere, it expands and cools.

_____ **4.** Evapotranspiration increases with increasing temperature.

_____ **5.** Most water used by industry is recycled.

_____ **6.** Irrigation is often necessary in areas having high evapotranspiration.

Choose the one best response. Write the letter of that choice in the space provided.

_____ **7.** Which of the following is an artificial means of producing fresh water from ocean water?

a. desalination
b. saltation
c. evapotranspiration
d. condensation

X
Groundwater

_____ **8.** Which of the following is represented by this diagram?

a. the earth's water budget
b. local water budget
c. groundwater movement
d. surface runoff

_____ **9.** The arrow labeled **X** represents:

a. absorption.
b. evaporation.
c. rejuvenation.
d. transpiration.

_____ **10.** Approximately what percentage of the earth's precipitation falls on the ocean?

a. 5%
b. 25%
c. 75%
d. 99%

Section 13.2

River Systems

Choose the one best response. Write the letter of that choice in the space provided.

_____ **1.** What is the term for the main stream and tributaries of a river?

a. river system
b. river bed
c. drainage basin
d. water gap

_____ **2.** The path that a river follows is called its:

a. tributary. **b.** gully. **c.** channel. **d.** meander.

_____ **3.** All of the sediment carried by a river is called the:

a. suspended load.
b. bed load.
c. stream load.
d. dissolved load.

_____ **4.** Channel erosion is quickest in rivers having a:

a. high volume of suspended load.
b. low volume of suspended load.
c. high volume of bed load.
d. low volume of bed load.

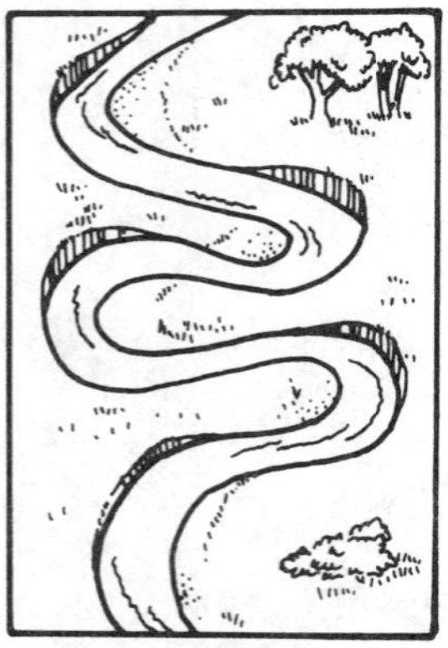

_____ **5.** The wide curves in the river in the diagram are:

a. meanders.
b. oxbows.
c. tributaries.
d. channels.

Complete each statement by writing the correct term or phrase in the space provided.

6. A river system begins to form when local precipitation exceeds ______________________ .

7. Drainage basins are also called ______________________ .

8. Capture of a river system by a river in a different watershed is called

______________________ .

9. Lengthening and branching of the upper end of a river is called ______________________ .

10. Water that cannot soak into the soil moves downslope as ______________________ .

Name ______________________ Class ______________ Date ______________

MODERN EARTH SCIENCE

Section 13.3

Stream Deposition

Read each statement below. If the statement is true, write *T* in the space provided. If the statement is false, write *F* in the space provided.

_____ **1.** Alluvial fan is another name for a delta.

_____ **2.** As the velocity of a stream decreases, its load of sediment also decreases.

_____ **3.** The volume of water in a stream remains constant from year to year.

_____ **4.** Artificial levees are resistant to river erosion.

_____ **5.** Human activities are often responsible for floods.

_____ **6.** Floodways often prevent streams from overflowing.

_____ **7.** Flooding is a natural stage in the development of a stream.

Complete each statement by writing the correct term or phrase in the space provided.

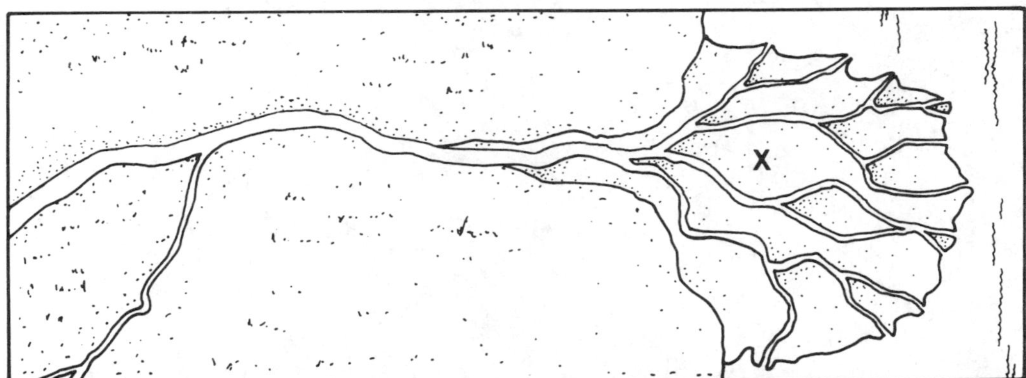

8. The region labeled **X** in the diagram is produced by stream ______________________.

9. The construction of a water dam usually results in the formation of an artificial

______________________.

10. One method of flood control is the building of artificial ______________________.

Name ____________________ Class ____________ Date ____________

Section 14.1

Water Beneath the Surface

Choose the one best response. Write the letter of that choice in the space provided.

_____ **1.** Which of the following factors most directly influences porosity?

a. grain composition
b. sorting of grains
c. grain size
d. permeability of grains

_____ **2.** Which of the following statements about the water table is true?

a. The water table generally follows the topography of the land surface.
b. The water table is found in the zone of aeration.
c. There is only one water table below all land surfaces.
d. The level of the water table remains constant.

_____ **3.** Capillary action in soil results in the upward movement of:

a. oxygen.
b. calcite.
c. carbon dioxide.
d. water.

_____ **4.** Contamination of coastal groundwater by salt water is primarily due to:

a. highway runoff in winter and spring.
b. lowering of the water table.
c. infiltration of toxic wastes.
d. introduction of hard water.

_____ **5.** When a sediment is well sorted, its particles are all about the same:

a. size. **b.** shape. **c.** weight. **d.** composition.

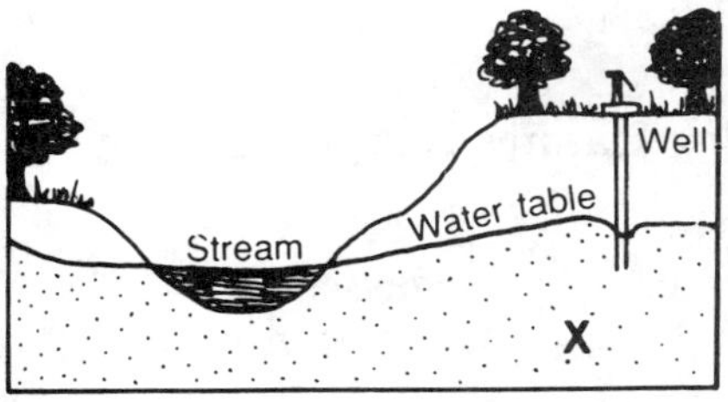

Well
Water table
Stream
X

_____ **6.** The groundwater region labeled **X** in the diagram represents a:

a. perched water table.
b. zone of saturation.
c. capillary fringe.
d. zone of aeration.

Complete each statement by writing the correct term or phrase in the space provided.

7. Water found below the earth's surface is called ____________________ .

8. The steepness of a water table's slope is known as its ____________________ .

9. The measure of how freely water passes through the open spaces in rock or sediment is referred to as ____________________ .

10. Large amounts of water can flow through and be stored in a body of rock called ____________________ .

Name ______________ Class ______________ Date ______________

Section 14.2

Wells and Springs

Read each statement below. If the statement is true, write *T* in the space provided. If the statement is false, write *F* in the space provided.

______ **1.** Because hot water tends to have fewer minerals dissolved in it than cold water, hot springs often contain very small concentrations of minerals.

______ **2.** The layer of cap rock in an artesian formation is impermeable.

______ **3.** Hot springs generally have temperatures at or very near the boiling point of water.

______ **4.** Geysers tend to build deposits of the silicate mineral geyserite.

______ **5.** Ordinary springs are commonly located where the water table intersects the ground surface.

Choose the one best response. Write the letter of that choice in the space provided.

______ **6.** Which of the following is often the source of water in a desert oasis?

a. an ordinary well **b.** a geyser
c. an artesian spring **d.** a hot spring

______ **7.** A sloping layer of permeable rock sandwiched between two layers of impermeable rock and exposed at the surface is:

a. a mud pot. **b.** a hot spring.
c. an ordinary well. **d.** an artesian formation.

Water

______ **8.** Which of the following features is represented by this diagram?

a. travertine terrace **b.** geyser
c. cap rock **d.** artesian well

______ **9.** Groundwater is sometimes heated beneath the earth's surface as it passes through areas where there has been recent:

a. severe drought. **b.** aquifer contamination.
c. mineral deposition. **d.** volcanic activity.

______ **10.** Pumping water from a well creates a lowered area of the water table called a:

a. cone of depression. **b.** fissure spring.
c. perched water table. **d.** paint pot.

Name ______________________ Class ______________ Date ____________

Section 14.3

Groundwater and Chemical Weathering

Choose the one best response. Write the letter of that choice in the space provided.

_____ **1.** The formations suspended from the roofs of caverns formed from the deposition of dissolved material are called:

a. fissures. **b.** natural bridges.
c. stalactites. **d.** disappearing streams.

_____ **2.** The mineral that forms cone-shaped cavern formations is:

a. calcite. **b.** magnesium. **c.** geyserite. **d.** iron.

_____ **3.** One of the major disadvantages of hard water is that it:

a. cannot be used for drinking.
b. inhibits the flow of water up through artesian wells.
c. contaminates groundwater aquifers.
d. damages appliances due to the buildup of mineral deposits.

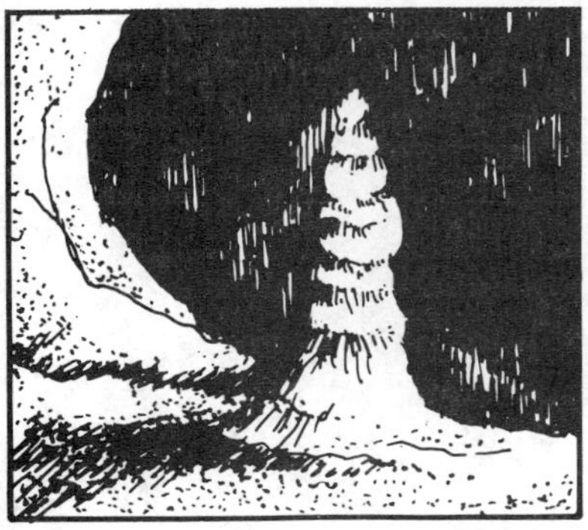

_____ **4.** What formation is represented in the diagram?

a. stalactite **b.** column
c. stalagmite **d.** natural bridge

Complete each statement by writing the correct term or phrase in the space provided.

5. A surface depression that forms as a result of the collapse of a cave roof is called a ______________________ .

6. Most caverns are formed when weak acids break down the rock through the process of ______________________ .

7. Water that contains few dissolved minerals is called ______________________ .

8. Regions where the effects of chemical weathering due to groundwater are clearly visible at the surface are said to have ______________________ .

9. Large tunnels or caves in areas with extensive limestone deposits that often contain smaller connecting chambers are called ______________________ .

10. The uncollapsed rock between a pair of sinkholes forms an arch of rock called a ______________________ .

Name ____________________ Class ____________ Date ____________

MODERN EARTH SCIENCE

Section 15.1

Glaciers: Moving Ice

Choose the one best response. Write the letter of that choice in the space provided.

_____ **1.** Solid rock over which a glacier has moved is often polished by the scraping action of:

a. firn. **b.** ice crystals.
c. erratics. **d.** rock particles.

_____ **2.** Long Island, New York is an example of:

a. an esker. **b.** a terminal moraine.
c. a kettle. **d.** a drumlin.

_____ **3.** Today, the Great Lakes drain through the:

a. Hudson River. **b.** Susquehanna River.
c. St. Lawrence River. **d.** Mohawk River.

_____ **4.** Which of the following features is represented in the diagram by number 1?

a. crevasses **b.** ice sheets
c. cirques **d.** icebergs

_____ **5.** An ice shelf is indicated by number:

a. 2. **b.** 3.
c. 4. **d.** 5.

Complete each statement by writing the correct term or phrase in the space provided.

6. The change from grainy firn to steel-blue ice is caused by ____________________ .

7. The rate of internal plastic flow in glaciers is determined by slope, ice temperature, and ice ____________________ .

8. Masses of moving ice that carve landscapes are called ____________________ .

9. The elevation above which ice and snow remain throughout the year is called the ____________________ .

10. The largest continental ice sheet in the world is located in ____________________ .

Section 15.2

Landforms Created by Glaciers

Read each statement below. If the statement is true, write *T* in the space provided. If the statement is false, write *F* in the space provided.

_____ **1.** Valley glaciers generally carve sharp, rugged features.

_____ **2.** Till is composed of sorted deposits of rock material.

_____ **3.** An esker is a ridge carved by a valley glacier.

_____ **4.** Most glacial lakes are formed by deposition rather than by erosion.

_____ **5.** Glaciers often form V-shaped valleys.

_____ **6.** Drumlins often occur in clusters.

Complete each statement by writing the correct term or phrase in the space provided.

GLACIAL DEPOSITS

	Description
1	parallel ridges of till at a glacier's leading edge
2	a ridge of unsorted material along the sides of a glacier
3	material that has been sorted and layered by streams of melted ice
4	a boulder that has been transported and deposited by a glacier
5	long, low mounds composed of sorted till

7. A lateral moraine is defined in the table by description number ______________________ .

8. A glacial erratic is defined in the table by description number ______________________ .

9. The scraping of solid rock by small rock particles embedded in a glacier is called

______________________ .

10. Salt lakes generally develop in areas that have high evaporation and low

______________________ .

Section 15.3

Ice Ages

Read each statement below. If the statement is true, write *T* in the space provided. If the statement is false, write *F* in the space provided.

_____ **1.** Melting of an ice sheet would result in a lowering of sea level.

_____ **2.** Ice ages may begin with a lowering of global temperatures.

_____ **3.** The distribution of solar energy reaching the earth depends on the earth's tilt.

_____ **4.** Most of the earth's ice ages have lasted about 1,000 years.

_____ **5.** During the last ice age, ice covered about half of the earth.

_____ **6.** All theories about the cause of ice ages involve changes in the amount of sunlight reaching the earth.

Complete each statement by writing the correct term or phrase in the space provided.

7. During ice ages, periods during which glaciers retreat are called ______________________ .

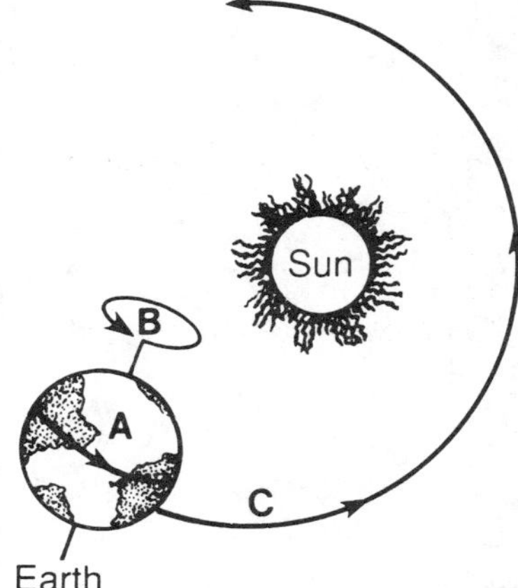

8. Precession is indicated in the diagram by the arrow labeled ______________________ .

9. A theory that ice ages are caused by changes in the earth's orbit and tilt was proposed by Milutin ______________________ .

10. A long period of climatic cooling during which ice covers much of the earth is called an ______________________ .

Section 16.1

Wind Erosion

Read each statement below. If the statement is true, write *T* in the space provided. If the statement is false, write *F* in the space provided.

_____ **1.** Desert pavement is hard-packed clay.

_____ **2.** Most sand grains are made of feldspar.

_____ **3.** Ventifacts are rocks that are smoothed by wind abrasion.

_____ **4.** All the material eroded by the wind is eventually deposited.

_____ **5.** Dunes usually form where soil is dry and unprotected.

Choose the one best response. Write the letter of that choice in the space provided.

_____ **6.** Which type of material is usually moved by saltation?

a. dust **b.** sand **c.** gravel **d.** cobbles

_____ **7.** The process in which fine, very dry soil particles are removed by wind is called:

a. deflation. **b.** emergence. **c.** hollowing. **d.** deposition.

_____ **8.** The downwind side of a dune is called the:

a. terrace. **b.** slipface. **c.** cliff. **d.** crescent.

_____ **9.** The fine-grained, yellowish sediment that is soft, easily eroded, and deposited by the wind is called:

a. silt. **b.** desert pavement.
c. loess. **d.** sedimentary rock.

Wind
Direction

_____ **10.** What type of dune is pictured in this diagram?

a. longitudinal
b. parabolic
c. transverse
d. barchan

Section 16.2

Wave Erosion

Choose the one best response. Write the letter of that choice in the space provided.

_____ **1.** Isolated columns of rock that result from ocean wave action on rock projections near shore are called sea:

a. cliffs. **b.** fractures. **c.** arches. **d.** stacks.

_____ **2.** A nearly level platform left just below the surface of the water as waves erode the shoreline back is called a:

a. terrace. **b.** berm. **c.** beach. **d.** bar.

_____ **3.** Which of the following is the source of the black sand at some beaches?

a. granite
b. volcanic rock
c. coral reefs
d. glacial deposits

_____ **4.** Sand deposits that connect islands are called:

a. reefs. **b.** cobbles. **c.** tombolos. **d.** spits.

_____ **5.** Projections of a shore are called:

a. estuaries. **b.** dunes. **c.** headlands. **d.** fiords.

Complete each statement by writing the correct term or phrase in the space provided.

6. When waves cut a hole completely through a shoreline rock projection, the formation that results is called a ______________ .

7. Extensions of wave-cut, flat surfaces formed by the deposition of eroded material some distance from shore are known as ______________ .

8. Long underwater ridges of sand that form offshore are called ______________ .

9. The term for the type of current that is created when waves approach a beach at the angle pictured in this diagram is ______________ .

10. The general direction in which sand will be moved along this beach is ______________ .

Name ______________ Class ______________ Date ______________

Section 16.3

Coastal Erosion and Deposition

Choose the one best response. Write the letter of that choice in the space provided.

_____ **1.** Sea level is now rising at an average rate of about:

a. 1 mm/yr. **b.** 5 mm/yr. **c.** 15 mm/yr. **d.** 25 mm/yr.

_____ **2.** The rising of the lithosphere can be the result of:

a. falling sea level. **b.** retreating glaciers.
c. eroding sea cliffs. **d.** newly forming atolls.

_____ **3.** One tenth of the shellfish-producing waters of the United States have been destroyed by:

a. erosion. **b.** pollution. **c.** overfishing. **d.** coral.

_____ **4.** In which direction will the barrier island in this diagram most likely migrate?

a. north **b.** south
c. east **d.** west

_____ **5.** The shallow water in the area labeled **X** is called a:

a. spit. **b.** fiord.
c. estuary. **d.** lagoon.

_____ **6.** What does coral extract from sea water in order to build its skeleton?

a. sodium chloride **b.** quartz
c. calcium carbonate **d.** feldspar

_____ **7.** Organisms that build coral reefs always live:

a. in cold, deep water. **b.** along emergent coastlines.
c. in warm, shallow water. **d.** along submergent coastlines.

Read each question and answer it in the space provided.

8. What is the term for the muddy or sandy parts of a lagoon that are visible at low tide? ______________

9. What is the term for the nearly circular reef that forms as a volcanic island disappears? ______________

10. What are narrow, deep bays with steep walls that result from the flooding of glacial valleys called? ______________

Name ______________________ Class ______________ Date ______________

Section 17.1

Determining Relative Age

Read each statement below. If the statement is true, write *T* in the space provided. If the statement is false, write *F* in the space provided.

_____ **1.** James Hutton stated that the geologic formations on the earth's crust took millions of years to develop.

_____ **2.** Disconformities are produced when folded or tilted rock layers undergo weathering.

_____ **3.** According to the law of crosscutting relationships, an intrusion is older than the rocks through which it cuts.

_____ **4.** Nonconformities occur between horizontal rock layers.

_____ **5.** The law of superposition could be used to determine the relative age of a sequence of igneous lava flows.

Complete each statement by writing the correct term or phrase in the space provided.

6. Layers in sedimentary rock are called ______________________ .

7. Small waves formed on the surface of sand by the action of water or wind are called ______________________ .

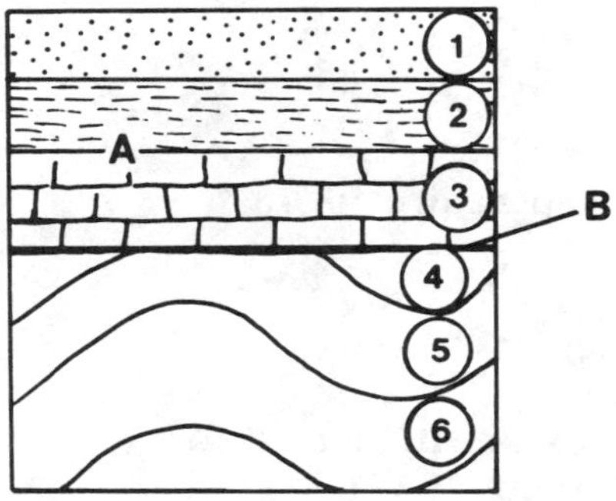

8. The boundary labeled **A** in the diagram represents a ______________________ .

9. The principle that can be used to determine the relative age of layers **1** through **3** in the diagram is the ______________________ .

10. Point **B** represents a type of rock feature called ______________________ .

Name ______________________ Class ______________ Date ______________

MODERN EARTH SCIENCE

Section 17.2

Determining Absolute Age

Choose the one best response. Write the letter of that choice in the space provided.

_____ **1.** Elements that emit atomic particles and energy are called:

a. radioactive. **b.** stable. **c.** magnetic. **d.** fluorescent.

_____ **2.** Nickel-63 has a half-life of 92 years. How much nickel-63 would remain in a 20 g sample at the end of 184 years?

a. 2 g **b.** 5 g **c.** 8 g **d.** 10 g

Cross Section of the Annual Deposition of Sediments in a Glacial Lake

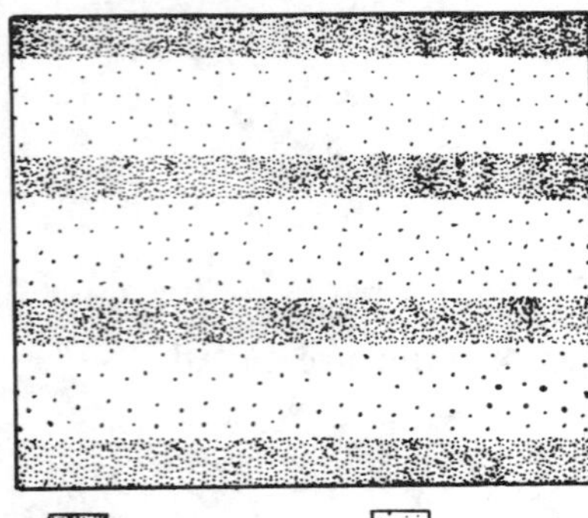

= clay = sand

_____ **3.** The sediment layers in the diagram are called:

a. molds. **b.** imprints.
c. casts. **d.** varves.

_____ **4.** How many years of deposition are probably represented in the diagram?

a. 3.5 **b.** 5
c. 6.5 **d.** 7

_____ **5.** The topmost layer in the diagram was most likely deposited during:

a. summer. **b.** spring.
c. winter. **d.** fall.

Read each question and answer it in the space provided.

6. What is the mass number of an element that contains 19 protons, 19 electrons, and 20 neutrons? ______________

7. Which radioactive element would most likely be used to date a wooden bowl believed to be about 10,000 years old? ______________

8. What two types of particles are emitted when uranium-238 decays? ______________

9. What did scientists measure to determine the approximate age of Niagara Falls? ______________

10. What is the average rate of sediment deposition for rocks such as limestone and sandstone? ______________

Name ______________________ Class ______________ Date ______________

Section 17.3

The Fossil Record

Read each statement below. If the statement is true, write *T* in the space provided. If the statement is false, write *F* in the space provided.

______ **1.** Geologists use index fossils to help locate oil and natural gas deposits.

______ **2.** An organism will only undergo mummification if it is buried by sediments at the bottom of a glacial lake.

______ **3.** Coprolites can be used to study the eating habits of ancient animals.

______ **4.** Entire animals can be fossilized in tar pits.

______ **5.** A frozen mastodon is an example of a trace fossil.

Choose the one best response. Write the letter of that choice in the space provided.

______ **6.** Paleontology is defined as the study of:

a. rock layers. **b.** fossils.
c. erosion. **d.** unconformities.

______ **7.** Fossils found exclusively in the rock layers of a particular geologic age are called:

a. index fossils. **b.** gastrolith fossils.
c. metamorphic fossils. **d.** complete fossils.

______ **8.** Which of the following is an example of a trace fossil?

a. an intact shark tooth **b.** a petrified log
c. a dinosaur footprint **d.** an insect in amber

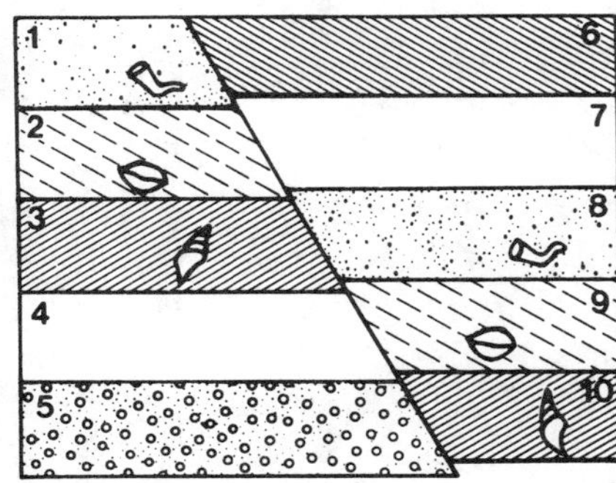

______ **9.** Which of the following rock layers in the diagram have the same relative age?

a. 1 and 6 **b. 2 and 9**
c. 3 and 7 **d. 5 and 10**

______ **10.** Approximately how old is a layer of rock containing ammonite fossils?

a. 245 to 550 million years **b.** 65 to 200 million years
c. 100 to 208 million years **d.** 10 to 65 million years

Name ______________________ Class ______________ Date ____________

Section 18.1

The Geologic Time Scale

Read each statement below. If the statement is true, write *T* in the space provided. If the statement is false, write *F* in the space provided.

_____ **1.** Scientists use events such as major changes in the earth's surface and climate and the extinction of various species as the basis for dividing the geologic time scale into units.

_____ **2.** Using the law of superposition and the study of index fossils, nineteenth-century scientists determined the relative ages of rock layers.

_____ **3.** Eras are subdivisions of periods.

_____ **4.** Scientists can use the geologic column to estimate the relative ages of rock layers only if the rock contains radioactive minerals.

_____ **5.** Fossils of mammals are common in Cenozoic rocks.

Complete each statement by writing the correct term or phrase in the space provided.

6. Rock layers in the geologic column can be distinguished from one another according to rock type and ______________________ .

7. The geologic era that followed Precambrian time is called the ______________________ .

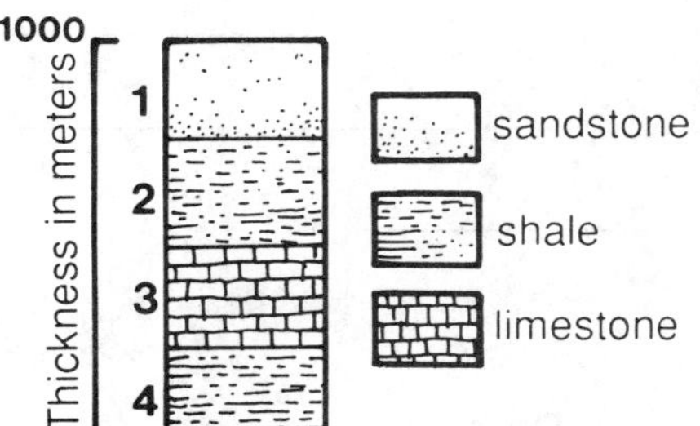

8. According to the geologic column in the diagram, the oldest rock layer is ______________________ .

9. Based on late Precambrian fossils and other types of geologic evidence, scientists theorize that life began ______________________ .

10. The era that is divided into epochs is the ______________________ .

Name ______________________ Class ______________ Date ______________

Section 18.2

Geologic History

Choose the one best response. Write the letter of that choice in the space provided.

_____ **1.** The presence of cross-bedded sandstone may indicate that the region where it is present was once:

a. an ocean. **b.** a mountain. **c.** a desert. **d.** a lake.

_____ **2.** Shields are large areas of exposed rock of what era?

a. Cenozoic **b.** Mesozoic **c.** Paleozoic **d.** Precambrian

_____ **3.** The fossil record indicates that which of the following thrived during the Paleozoic Era?

a. reptiles
b. advanced forms of marine life
c. mammals
d. marine algae and bacteria

_____ **4.** The Devonian Period is also known as the Age of:

a. Birds. **b.** Lizards. **c.** Mammals. **d.** Fishes.

_____ **5.** Dinosaurs first appeared during the:

a. Triassic Period.
b. Pliocene Epoch.
c. Cretaceous Period.
d. Oligocene Epoch.

Read each question and answer it in the space provided.

6. During which era did glaciers cover nearly one third of the earth's land area? ______________

7. During which epoch did the first modern horses appear? ______________

The Paleozoic Era

Cambrian → Ordovician → _______ → Devonian

TIME →

8. Which geologic period is missing from this diagram? ______________

9. What are the most common Precambrian fossils called? ______________

10. Which period of the Cenozoic Era includes the time before the last major ice age? ______________

Section 19.1

Movements of the Continents

Choose the one best response. Write the letter of that choice in the space provided.

_____ **1.** From which continent did India break away to move northward?

a. Eurasia
b. South America
c. Africa
d. Antarctica

_____ **2.** Approximately how far do crustal plates move in a year?

a. a few millimeters
b. a few meters
c. a few centimeters
d. a few kilometers

_____ **3.** The continents existing today contain large areas of deformed Precambrian rocks called:

a. cratons.
b. asthenospheres.
c. shields.
d. lithospheres.

The Breakup of Pangaea	
1	closing of the sea in the east
2	opening of an east-west rift
3	opening of the sea in the west
4	opening of a north-south rift
5	clockwise rotation of the northern landmass

_____ **4.** According to the table, which event caused the formation of the South Atlantic Ocean?

a. event 1
b. event 3
c. event 4
d. event 5

_____ **5.** Event 5 led to the formation of:

a. the North Atlantic Ocean.
b. Panthalassa.
c. the South Atlantic Ocean.
d. the Tethys Sea.

Read each question and answer it in the space provided.

6. After breaking away from Gondwanaland, India collided with what modern continent? ________________

7. What is the name for the movement of the earth's crustal plates? ________________

8. What scientist thought that Pangaea was the continental mass from which the modern continents split off? ________________

9. What landmass split from the supercontinent to form South America and Australia? ________________

10. The modern Mediterranean Sea formed as a result of the closing off of what ancient body of water? ________________

Name ______________________ Class ______________ Date ______________

M O D E R N   E A R T H   S C I E N C E

Section 19.2

Growth of a Continent: North America

Read each statement below. If the statement is true, write *T* in the space provided. If the statement is false, write *F* in the space provided.

_____ **1.** Precambrian rocks are exposed at the top of the Grand Canyon.

_____ **2.** The ancient landmass around which the present North American continent developed first took shape during the Mesozoic Era.

_____ **3.** Much of North America was once covered by an inland sea.

_____ **4.** Much of what is now the Atlantic coastal area of North America was once attached to the African plate.

_____ **5.** The tectonic activity that accompanied the breakup of Pangaea played a large role in shaping North America.

Choose the one best response. Write the letter of that choice in the space provided.

_____ **6.** During the Cenozoic Era, violent movements along plate boundaries in the western part of North America caused the uplift of the:

a. Sierra Nevada. **b.** Rocky Mountains.
c. Colorado Plateau. **d.** Great Plains.

_____ **7.** Terranes are best described as:

a. tilted fault-blocks.
b. exposed portions of cratons.
c. pieces of oceanic crust that are added to continents.
d. sediment-filled troughs.

_____ **8.** The Paleozoic Era ended approximately how many millions of years ago?

a. 408 **b.** 245 **c.** 144 **d.** 65

Features of the North American Continent	
1	Appalachian Mountains
2	sea covering western geosyncline
3	shallow inland sea
4	Rocky Mountains

_____ **9.** During which geologic time period did North America have the features listed in the table?

a. Precambrian **b.** Paleozoic
c. Mesozoic **d.** Cenozoic

_____ **10.** What feature produced the sediments that eventually formed the Great Plains?

a. 1 **b.** 2
c. 3 **d.** 4

Section 19.3

Formation of the Grand Canyon

Read each statement below. If the statement is true, write *T* in the space provided. If the statement is false, write *F* in the space provided.

_____ **1.** An outcrop is an exposed rock layer of any type.

_____ **2.** The Grand Canyon is an artificial outcrop of sedimentary rock.

_____ **3.** The Permian Period is not represented by rock layers in the Grand Canyon.

_____ **4.** Radioactive dating is used to estimate the relative age of fossils.

_____ **5.** Limestone layers in the Grand Canyon often contain fossils of marine organisms.

Read each question and answer it in the space provided.

6. What river cut the gorge that forms the Grand Canyon? ____________

7. What is the name for the uplifted region that includes the Grand Canyon? ____________

8. The lack of deposits from the Ordovician Period in the Grand Canyon is due to what process? ____________

Rock Layers of the Grand Canyon

Time Period	Rock Layers Formed
A	Kaibab limestone Toroweap formation Coconino sandstone **1** Supai formation
B	Redwall limestone
C	Muav limestone **2** Tapeats sandstone Vishnu schist

9. What geologic time period is represented by **C**? ____________

10. What rock type is represented by numbers **1** and **2**? ____________

Section 20.1

The Water Planet

Read each statement below. If the statement is true, write *T* in the space provided. If the statement is false, write *F* in the space provided.

_____ **1.** A bathyscaph is a submersible that is self-propelled and free-moving.

_____ **2.** The Pacific Ocean is the largest of the earth's principal oceans.

_____ **3.** The earth is the only known planet with a covering of liquid water.

_____ **4.** The *Alvin* is an example of a bathysphere.

_____ **5.** When the remains of the *Titanic* were discovered, many of the photographs of the inside of the ship were taken with a robot craft.

Complete each statement by writing the correct term or phrase in the space provided.

6. The second largest of the earth's principal oceans is the ________________ .

7. The general name for an underwater research vessel is a ________________ .

8. The technique in which sound waves are used to measure depth is called

________________ .

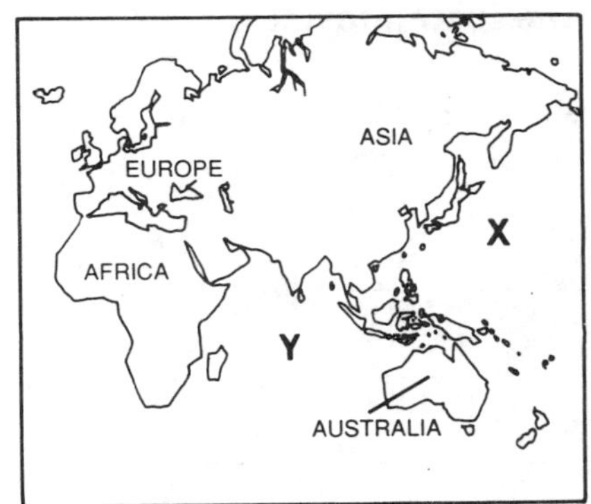

9. The area labeled **X** in the diagram is part of the

________________ .

10. The ocean labeled **Y** is the ________________ .

Name ______________________ Class ______________ Date ______________

MODERN EARTH SCIENCE

Section 20.2

Features of the Ocean Floor

Choose the one best response. Write the letter of that choice in the space provided.

_____ **1.** The boundary between continental crust and oceanic crust occurs at the base of the:

a. continental shelf. **b.** abyssal plain.
c. continental slope. **d.** mid-ocean ridge.

_____ **2.** Which ocean contains the greatest number of deep-sea trenches?

a. Atlantic **b.** Pacific **c.** Indian **d.** Arctic

_____ **3.** The shallowest portion of the continental margin is the:

a. continental shelf. **b.** mid-ocean ridge.
c. continental rise. **d.** abyssal plain.

_____ **4.** Which of the following describes a seamount?

a. ocean floor at the edge of a continental margin
b. underwater mountain range
c. sediment piled at the base of the continental slope
d. isolated mid-ocean volcano

_____ **5.** Which feature shown in the diagram is more extensive in the Atlantic Ocean than in the Pacific Ocean?

a. 1 **b. 2**
c. 3 **d. 4**

Complete each statement by writing the correct term or phrase in the space provided.

6. The portion of the ocean floor made up of oceanic crust is called the

______________________ .

7. Deep valleys in the continental slope are called ______________________ .

8. The most level part of the ocean floor is called the ______________________ .

9. The part of a continent that is covered by the ocean is the ______________________ .

10. The boundary between continental crust and oceanic crust at an inactive margin is the

______________________ .

Name ______________________ Class ______________ Date ______________

M O D E R N E A R T H S C I E N C E

Section 20.3

Ocean-Floor Sediments

Read each statement below. If the statement is true, write *T* in the space provided. If the statement is false, write *F* in the space provided.

_____ **1.** Most deep-ocean sediments come from continental erosion.

_____ **2.** Silica and calcium carbonate are commonly found in organic ocean sediments.

_____ **3.** Calcareous ooze is found at the ocean's greatest depths.

_____ **4.** Meteorites contribute to ocean sediments.

_____ **5.** Icebergs commonly deposit ocean sediments on land.

_____ **6.** Turbidity currents are caused by winds and temperature differences.

Complete each statement by writing the correct term or phrase in the space provided.

7. Ooze made up mostly of silicon dioxide is called ______________________ .

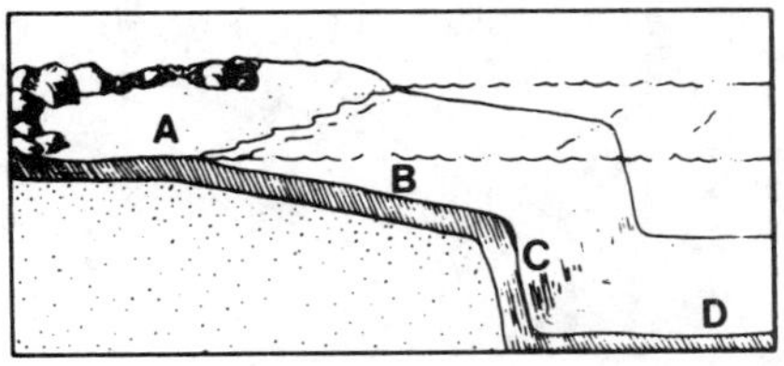

8. In the diagram, the finest sediments are most likely to be found at the point labeled ______________________ .

9. Potato-shaped lumps of minerals on the ocean floor are called ______________________ .

10. Red clay and ooze are examples of the general type of sediment called ______________________ .

Name ______________________ Class ______________ Date ______________

Section 21.1

Properties of Ocean Water

Read each statement below. If the statement is true, write *T* in the space provided. If the statement is false, write *F* in the space provided.

_____ **1.** Ocean water is 78 percent pure water.

_____ **2.** The salinity of ocean water tends to be lower near fresh water runoff from rivers.

_____ **3.** The total amount of solar energy falling on the surface of the ocean is much greater at the poles than at the equator.

_____ **4.** The amount of dissolved solids in the water affects the density of ocean water.

_____ **5.** The relative amounts of dissolved substances in ocean water have remained almost constant over millions of years.

Complete each statement by writing the correct term or phrase in the space provided.

6. Of the three principal gases dissolved in ocean water, the one that dissolves most easily is

______________________ .

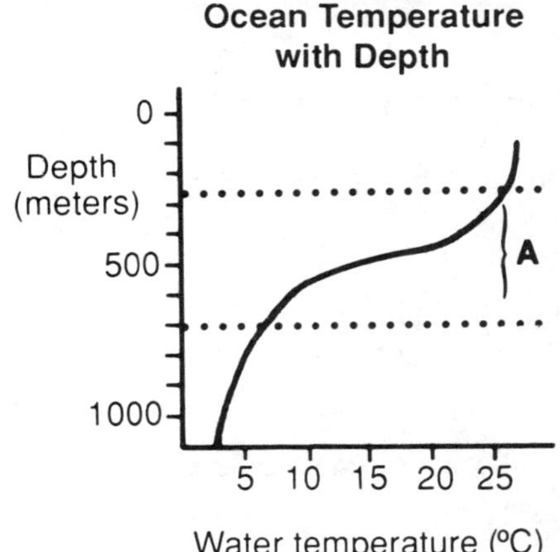

7. The region labeled **A** in the diagram indicates a zone called the ______________________ .

8. The layer of floating ice that covers the ocean surface near the poles is called

______________________ .

9. The most abundant dissolved elements in ocean water are sodium and

______________________ .

10. Clear water generally appears blue because blue wavelengths are the only wavelengths that tend to be ______________________ .

Name ______________________ Class ____________ Date ____________

Section 21.2

Life in the Oceans

Choose the one best response. Write the letter of that choice in the space provided.

_____ 1. Nearly all ocean life is regulated by the:

a. life processes of plants.
b. type of sediment on the ocean floor.
c. addition of substances to ocean water.
d. abundance of swimming ocean animals.

_____ 2. The process by which deep, nutrient-rich water moves upward to replace surface water that has blown farther offshore is called:

a. thermocline. **b.** distilling. **c.** desalination. **d.** upwelling.

_____ 3. Plant growth in the ocean is restricted to a depth that extends no more than:

a. 50 m below the surface. **b.** 100 m below the surface.
c. 200 m below the surface. **d.** 300 m below the surface.

_____ 4. Which benthic environment begins at the end of the continental slope?

a. sublittoral **b.** intertidal **c.** neritic **d.** bathyal

_____ 5. Large ocean animals that swim and eat microscopic plants and animals are:

a. nekton. **b.** zooplankton. **c.** aquaculture. **d.** anemones.

Complete each statement by writing the correct term or phrase in the space provided.

6. The general name for free-floating microscopic plants and animals is

______________________ .

7. The process by which microscopic plants produce food from sunlight is

______________________ .

8. Organisms that live on the ocean floor, such as oysters and sea stars, are examples of the general group of organisms called ______________________ .

Sea level
X
Continental shelf

9. The pelagic zone labeled **X** in the diagram is the

______________________ .

10. The benthic zone characterized by exposure at low tides and breaking waves is the ______________________ .

Section 21.3

Ocean Resources

Read each statement below. If the statement is true, write *T* in the space provided. If the statement is false, write *F* in the space provided.

_____ **1.** Evaporating ocean water to desalinate it requires very little heat energy.

_____ **2.** When ocean water freezes, the first ice crystals that form are free of salt.

_____ **3.** Traces of the insecticide DDT have been detected in open ocean.

_____ **4.** Oil spills from offshore drilling are harmless to marine organisms and sea birds.

_____ **5.** An ocean farm could potentially produce more protein-rich food than a land farm of the same size.

Choose the one best response. Write the letter of that choice in the space provided.

_____ **6.** The method used to remove salt by evaporating ocean water is called:

a. freezing.
b. reverse osmosis.
c. distillation.
d. electrolysis.

_____ **7.** Manganese from nodules in the ocean is used to manufacture certain types of:

a. plastic. **b.** steel. **c.** fertilizer. **d.** glass.

_____ **8.** Which of the following is the most valuable resource currently taken from the ocean?

a. petroleum **b.** iron **c.** calcium **d.** sodium

_____ **9.** What term is used to describe the practice of developing and raising special breeds of marine animals and plants for food?

a. distillation **b.** ocean fertilization
c. aquaculture **d.** artificial upwelling

_____ **10.** What part of the ocean is in the greatest danger of pollution?

a. polar waters **b.** coastal waters
c. deep marine waters **d.** intermediate ocean depths

Name ______________________ Class ______________ Date ______________

Section 22.1

Ocean Currents

Choose the one best response. Write the letter of that choice in the space provided.

_____ **1.** Which type of current results from underwater landslides?

a. rip current **b.** turbidity current
c. longshore current **d.** tidal current

_____ **2.** Which of the following best describes the water found in the deep currents of the Atlantic Ocean?

a. cold with high salinity **b.** cold with low salinity
c. warm with high salinity **d.** warm with low salinity

_____ **3.** The currents in the northern Indian Ocean are governed by winds called:

a. monsoons. **b.** tsunamis. **c.** trade winds. **d.** tombolos.

_____ **4.** Deep Atlantic currents are formed largely as a result of the:

a. erosion by surface winds. **b.** heating of surface water.
c. rotation of the earth. **d.** sinking of denser water.

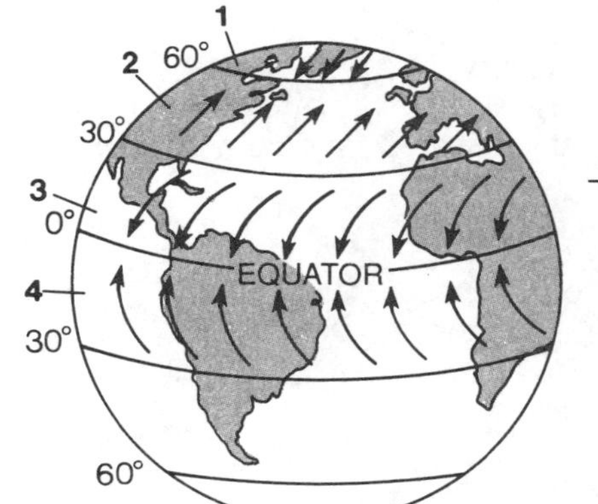

_____ **5.** Which number in the diagram represents the location of the winds called the westerlies?

a. 1 **b.** 2
c. 3 **d.** 4

Read each question and answer it in the space provided.

6. From which direction do the trade winds blow in the Northern Hemisphere? ______________

7. What is the deflection of the earth's winds and ocean currents caused by the earth's rotation called? ______________

8. What is the name of the calm area of the North Atlantic Ocean that is completely encircled by surface currents, trapping debris within the area? ______________

9. Which warm-water current is found just off the eastern coast of the United States? ______________

10. What deep-water current travels north along the ocean floor for thousands of kilometers? ______________

Section 22.2

Ocean Waves

Read each statement below. If the statement is true, write *T* in the space provided. If the statement is false, write *F* in the space provided.

_____ **1.** An undertow is generally a very strong current.

_____ **2.** Energy is transferred from wind to water to form waves.

_____ **3.** The depth of the water relative to wave length determines when a wave breaks.

_____ **4.** If the ocean shore slopes gently, waves will generally break with great force.

_____ **5.** The lowest point between two wave crests is a trough.

Choose the one best response. Write the letter of that choice in the space provided.

_____ **6.** The distance that wind can blow across open water is called the:

a. wavelength. **b.** fetch. **c.** wave period. **d.** swell.

_____ **7.** Which of the following factors influences the size of a wave?

a. wind speed
b. air temperature
c. deep-current speed
d. water density

_____ **8.** When water returns to the ocean through channels in underwater sandbars, it may produce a current known as a:

a. turbidity current.
b. bottom current.
c. rip current.
d. surface current.

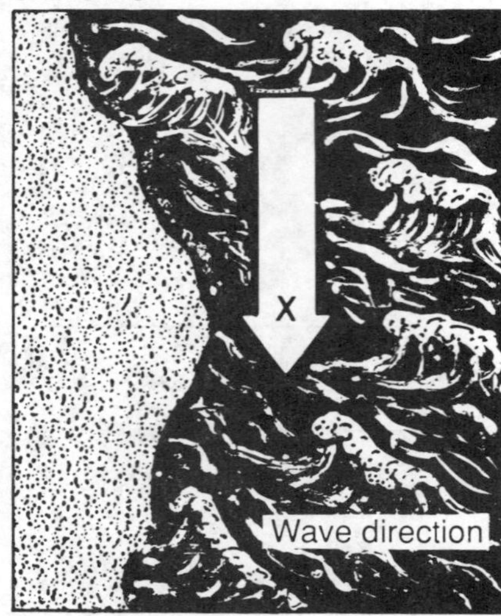

_____ **9.** The movement of water in the direction of arrow **X** in this diagram represents:

a. a tidal current.
b. an undertow.
c. a longshore current.
d. a breaker.

_____ **10.** What is calculated by dividing the wavelength of a wave by the wave's period?

a. wave speed
b. size of wave crest
c. wave height
d. size of wave trough

Name ______________________ Class ____________ Date ____________

Section 22.3

Tides

Read each statement below. If the statement is true, write *T* in the space provided. If the statement is false, write *F* in the space provided.

______ **1.** The position of the sun has the greatest influence on the tides.

______ **2.** High tides occur at the same time each day.

______ **3.** According to Newton's law of gravitation, the gravitational pull of the moon on the earth and its waters causes the rise and fall of the tides.

______ **4.** Tidal range is the difference in time between two successive high tides.

______ **5.** Tidal patterns are greatly influenced by the size, shape, and location of the ocean basin in which they occur.

______ **6.** There are always two tidal bulges somewhere on the earth's surface.

Read each question and answer it in the space provided.

7. What is the term for a tidal current that flows toward the coast? ______________

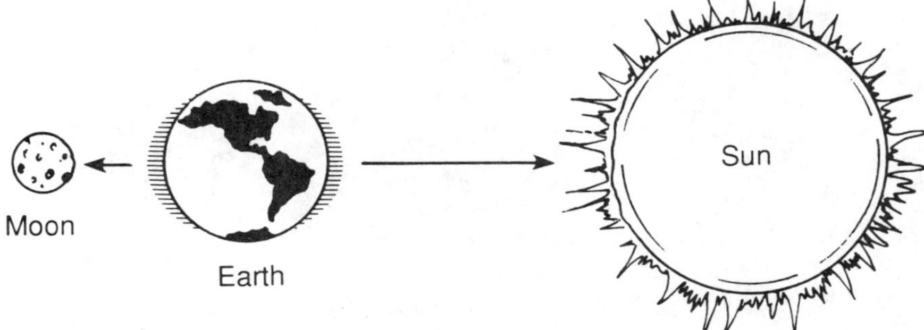

8. What type of tide is produced when the earth, sun, and moon are aligned as in this diagram? ______________

9. What is the term for the surge of water that is sometimes seen when tidal currents move up rivers through a long bay? ______________

10. What is the term for the tidal bulge that occurs during the first and third quarters of the moon? ______________

Name ______________________ Class ______________ Date ______________

MODERN EARTH SCIENCE

Section 23.1

Characteristics of the Atmosphere

Choose the one best response. Write the letter of that choice in the space provided.

_____ **1.** In which layer of the earth's atmosphere is most ozone found?

a. troposphere **b.** thermosphere **c.** stratosphere **d.** mesosphere

_____ **2.** Which of the following is a unit used to measure atmospheric pressure?

a. millibar **b.** newton **c.** milligram **d.** degree

_____ **3.** Water vapor is added to the atmosphere primarily by:

a. convection. **b.** evaporation.
c. condensation. **d.** precipitation.

_____ **4.** Acid rain usually results when water vapor in the atmosphere combines with:

a. ash blown from forest fires.
b. products formed from the breakdown of ozone.
c. gases from aerosol cans.
d. gases emitted by the burning of fossil fuels.

_____ **5.** Which of the following meteorological instruments does this illustration represent?

a. altimeter
b. mercurial barometer
c. thermometer
d. aneroid barometer

Complete each statement by writing the correct term or phrase in the space provided.

6. The most abundant compounds in the earth's atmosphere are water vapor and

______________________ .

7. The ratio of the weight of air to the area on which it presses is called

______________________ .

8. The layer of the atmosphere closest to the earth's surface is the ______________________ .

9. Any substance in the atmosphere that is harmful to organisms or property is called

______________________ .

10. The general condition of the atmosphere at a particular time and place is called

______________________ .

Name ______________________ Class ____________ Date ____________

MODERN EARTH SCIENCE

Section 23.2

Solar Energy and the Atmosphere

Read each statement below. If the statement is true, write *T* in the space provided. If the statement is false, write *F* in the space provided.

_____ **1.** Almost all of the solar energy reaching the earth is reflected back into space.

_____ **2.** Air is a good conductor of heat.

_____ **3.** Radiation from the sun travels through space in the form of waves.

_____ **4.** Desert temperatures usually show large changes between day and night.

_____ **5.** Increases in atmospheric levels of carbon dioxide would probably cause the atmosphere to become warmer.

Complete each statement by writing the correct term or phrase in the space provided.

6. The fraction of solar radiation reflected by a surface is called the surface's

_________________________ .

7. The sun appears red at sunset due to the process called _________________________ .

8. The movement of air due to uneven heating is called _________________________ .

9. The name of the process by which the atmosphere traps infrared rays is the

_________________________ .

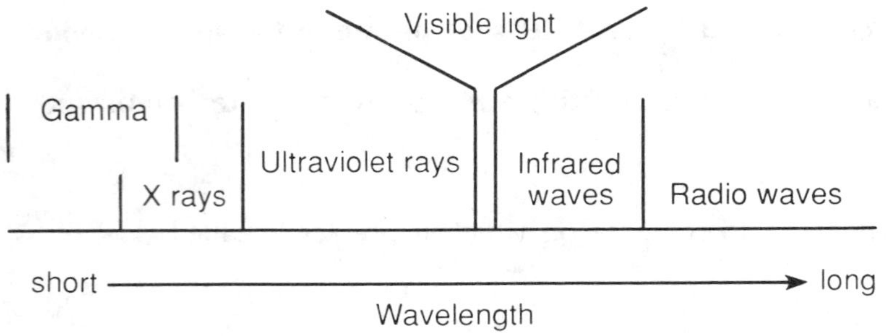

Visible light
Gamma
X rays
Ultraviolet rays
Infrared waves
Radio waves
short → long
Wavelength

10. The entire range of radiation represented by this diagram is called the

_________________________ .

Name ______________________ Class ______________ Date ____________

MODERN EARTH SCIENCE

Section 23.3

Winds

Read each statement below. If the statement is true, write *T* in the space provided. If the statement is false, write *F* in the space provided.

_____ **1.** A sea breeze generally occurs at night.

_____ **2.** Polar easterlies are generally strong winds.

_____ **3.** The positions of the global wind belts shift during the year.

_____ **4.** In the Northern Hemisphere, the Coriolis effect deflects surface winds to the left.

_____ **5.** A mountain breeze is created when cooler air descends from mountain peaks.

_____ **6.** Areas close to the equator have low atmospheric pressure due to ascending air.

Choose the one best response. Write the letter of that choice in the space provided.

_____ **7.** How many convection cells are in the Southern Hemisphere?

a. 3 **b.** 4 **c.** 6 **d.** 8

_____ **8.** Near the equator, the narrow zone formed by the convergence of the trade winds is called the:

a. horse latitudes. **b.** jet stream.
c. prevailing westerlies. **d.** doldrums.

_____ **9.** The general direction of flow of the jet stream is toward the:

a. north. **b.** south. **c.** east. **d.** west.

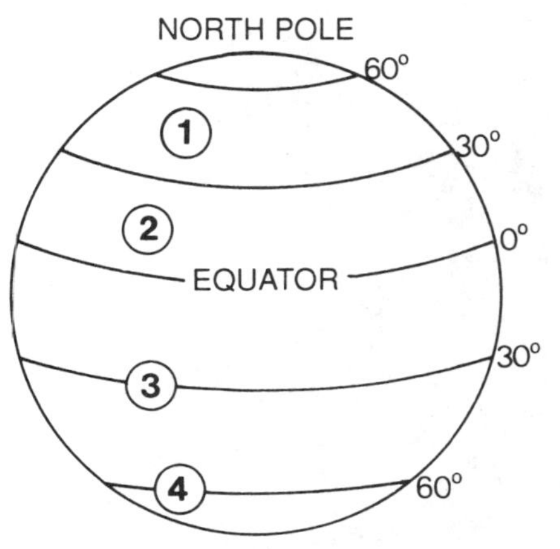

_____ **10.** Which number represents the location of the subpolar lows?

a. 1 **b.** 2
c. 3 **d.** 4

Section 24.1

Atmospheric Moisture

Read each statement below. If the statement is true, write *T* in the space provided. If the statement is false, write *F* in the space provided.

______ **1.** Sublimation is the process by which a solid changes to a liquid.

______ **2.** Humid air usually has a higher dew point than dry air.

______ **3.** Water usually condenses when air temperature falls below dew point.

______ **4.** The amount of water vapor that a volume of air can hold decreases as air temperature rises.

______ **5.** The heat energy necessary for evaporation of ocean water is absorbed from the sun.

Read each question and answer it in the space provided.

6. What is the name of the instrument that uses wet-bulb and dry-bulb thermometers to measure relative humidity? ______________

7. In what form is most atmospheric water found? ______________

Sample	T °C	Mass of Water Vapor	Mass of Water Vapor at Saturation
1	20°	12 g	17 g
2	32°	14 g	34 g
3	41°	20 g	56 g

8. What is the relative humidity of air sample **2** in the table? ______________

9. What is the principal source of atmospheric moisture? ______________

10. What forms on surfaces when the dew point is below the freezing temperature of water? ______________

Section 24.2

Clouds and Fog

Choose the one best response. Write the letter of that choice in the space provided.

_____ **1.** Which type of cloud usually forms in fair weather?

a. nimbostratus **b.** cumulus **c.** cirrocumulus **d.** nimbus

_____ **2.** Which type of fog occurs as warm, moist air moves across a cold surface?

a. radiation **b.** upslope **c.** advection **d.** steam

_____ **3.** What type of cloud forms when warm, moist air overrides a layer of cool air?

a. stratus **b.** cumulonimbus
c. cirrus **d.** altonimbus

_____ **4.** Forceful lifting of an air mass can occur as the air meets a:

a. body of water. **b.** mountain range.
c. cloud bank. **d.** valley.

_____ **5.** Which type of cooling is most likely to occur as the air mass in the diagram moves over the water?

a. condensational **b.** adiabatic
c. convective **d.** advective

Read each question and answer it in the space provided.

6. What level is marked by the base of a cloud layer? ______________

7. What type of fog usually forms over inland rivers and lakes? ______________

8. What are temperature changes that result solely from the expansion and compression of air called? ______________

9. What type of cloud creates a halo effect around the sun? ______________

10. What is another name for cumulonimbus clouds? ______________

Section 24.3

Precipitation

Read each statement below. If the statement is true, write *T* in the space provided. If the statement is false, write *F* in the space provided.

______ **1.** Raindrops are generally larger than 5 mm in diameter.

______ **2.** Silver iodide is used in cloud seeding because it resembles ice crystals.

______ **3.** Cloud droplets must increase in size in order to fall as precipitation.

______ **4.** Cloud seeding may eventually be used to control the severity of storms.

______ **5.** Hail is formed by rain that freezes as it strikes the ground.

Choose the one best response. Write the letter of that choice in the space provided.

______ **6.** In which type of cloud is hail usually formed?

a. cumulonimbus
b. cirrostratus
c. nimbostratus
d. cirrocumulus

______ **7.** The collision and combination of large and small cloud droplets are described as:

a. supercooling.
b. coalescence.
c. precipitation.
d. convection.

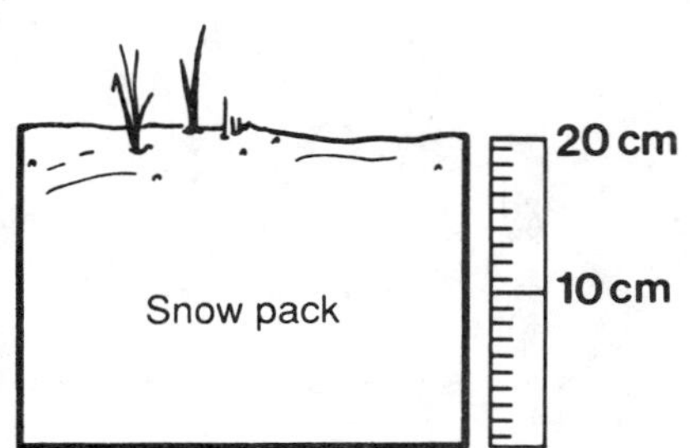

______ **8.** How much liquid water would most likely result from the melting of snow in the diagram?

a. 1 cm
b. 2 cm
c. 10 cm
d. 20 cm

______ **9.** Which type of precipitation forms as rain falls through a layer of freezing air?

a. rain
b. snow
c. drizzle
d. sleet

______ **10.** Snow forms in supercooled clouds by the:

a. combination of supercooled droplets.
b. reaction between dry ice and water.
c. condensation of water onto ice crystals.
d. aggregation of ice crystals.

Name ______________________ Class ____________ Date ____________

MODERN EARTH SCIENCE

Section 25.1

Air Masses

Read each statement below. If the statement is true, write *T* in the space provided. If the statement is false, write *F* in the space provided.

_____ **1.** An air mass can be thousands of kilometers in diameter.

_____ **2.** The temperature and humidity of an air mass depend primarily on its source region.

_____ **3.** An air mass that forms over Mexico is classified as an mP air mass.

_____ **4.** In summer, maritime polar Pacific air masses bring cool, foggy weather to the Pacific Coast of the United States.

_____ **5.** Frequent variations in air pressure cause air masses to become stationary.

Choose the one best response. Write the letter of that choice in the space provided.

_____ **6.** Which letter designation is used for an air mass that forms over the southwestern United States?

a. mT **b.** mP **c.** cP **d.** cT

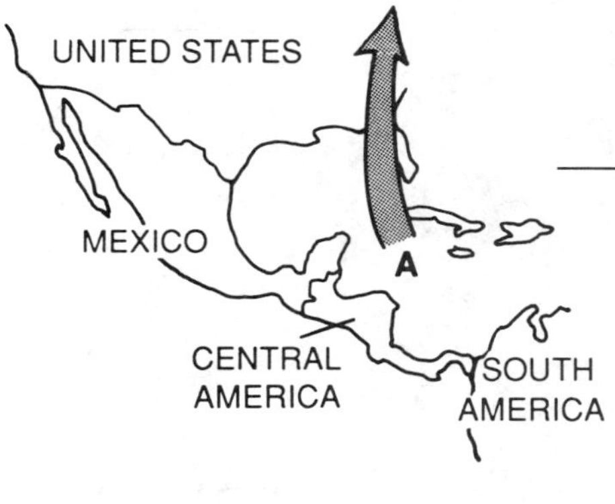

UNITED STATES
MEXICO
A
CENTRAL AMERICA
SOUTH AMERICA

_____ **7.** What type of air mass usually forms in Region **A** and moves along the path of the arrow on the map?

a. continental tropical Mexican
b. maritime tropical gulf
c. maritime tropical Atlantic
d. continental tropical desert

_____ **8.** Which area below is a source region for mT air masses?

a. southwestern United States
b. northern Canada
c. New England coast
d. southern Pacific Ocean

_____ **9.** Which type of air mass brings cool, dry summer weather to the middle of the United States?

a. maritime tropical gulf
b. maritime polar Pacific
c. continental tropical
d. maritime polar Atlantic

_____ **10.** During the summer, air masses that form over the southwestern United States usually bring weather that is:

a. hot and clear.
b. rainy and cold.
c. cool and clear.
d. hot and humid.

Name ______________________ Class ______________ Date ______________

MODERN EARTH SCIENCE

Section 25.2

Fronts

Read each question and answer it in the space provided.

1. What type of storm is usually found along a squall line? ______________
2. What type of front is generally preceded by cirrus and cirrostratus clouds? ______________
3. What is the name of the boundary at which cold polar air meets the warmer air of the middle latitudes? ______________
4. What type of front forms in the first stage of a wave cyclone? ______________
5. What causes the clockwise circulation of air around a center of high pressure in the Northern Hemisphere? ______________
6. What are hurricanes that form over the western Pacific Ocean called? ______________
7. What is the name for a small storm that forms when a thunderstorm meets high-altitude horizontal winds? ______________
8. What term is used to describe the center of a hurricane? ______________
9. What type of front is produced when a cold front overtakes and lifts warm air completely off the ground? ______________

Read the statement and answer it in the space provided.

10. Explain how hurricanes form.

____

____

____

____

____

____

Name ______________________ Class ______________ Date ______________

Section 25.3

Weather Instruments

Choose the one best response. Write the letter of that choice in the space provided.

_____ **1.** When a weather map shows a wind symbol at 270°, it indicates that the wind is coming from the:

a. north. **b.** south. **c.** east. **d.** west.

_____ **2.** A barometer is used to measure:

a. wind speed.
b. precipitation rate.
c. humidity level.
d. air pressure.

°C
100
50
0

_____ **3.** What type of thermometer is shown in the diagram?

a. bimetal
b. liquid
c. thermograph
d. electrical

_____ **4.** Satellite images made by using infrared energy primarily reveal information about atmospheric:

a. temperature. **b.** humidity. **c.** pollution. **d.** wind speed.

_____ **5.** Radiosondes are most often used to measure:

a. cloud droplet size.
b. aerosol levels.
c. ion formation rates.
d. relative humidity.

Read each question and answer it in the space provided.

6. What two temperature scales are commonly used in the United States? ______________

7. A wind vane points north on a very windy day. In directional terms, how is the wind described? ______________

8. Which instrument do scientists use to measure wind speed? ______________

9. What information do scientists gather by tracking the path of a helium-filled balloon? ______________

10. What do meteorologists measure by using a psychrometer? ______________

Name ______________________ Class ______________ Date ______________

MODERN EARTH SCIENCE

Section 25.4

Forecasting the Weather

Read each statement below. If the statement is true, write *T* in the space provided. If the statement is false, write *F* in the space provided.

_____ **1.** The World Meteorological Organization was created to promote the rapid exchange of weather information.

_____ **2.** Station models summarize weather conditions at individual weather observation stations.

_____ **3.** The dew point is a measure of air pressure.

_____ **4.** Extended forecasts usually predict precipitation very accurately.

_____ **5.** Hurricanes have been successfully controlled using cloud seeding.

Read each question and answer it in the space provided.

6. On a weather map, what is a line connecting the points of equal atmospheric pressure called? _______________

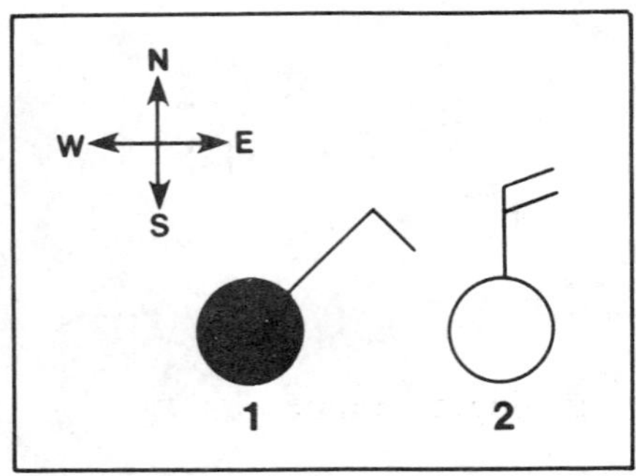

7. What is the wind direction indicated by the symbol labeled **2**? _______________

8. What cloud conditions are indicated by the symbol labeled **1**? _______________

9. What do scientists try to produce by adding freezing nuclei to supercooled clouds? _______________

10. Daily forecasts usually predict weather conditions for how many consecutive hours? _______________

Name ______________________ Class ______________ Date ______________

MODERN EARTH SCIENCE

Section 26.1

Factors That Affect Climate

Read each statement below. If the statement is true, write *T* in the space provided. If the statement is false, write *F* in the space provided.

_____ **1.** The specific heat of land is higher than that of water.

_____ **2.** Winds that blow down mountain slopes may be either warm or cold.

_____ **3.** At the equator, both days and nights are about 12 hours long in all seasons.

_____ **4.** Wave cyclones frequently develop in the subtropical highs.

_____ **5.** The Gulf Stream has a moderating effect on the climate of the east coast of North America.

Choose the one best response. Write the letter of that choice in the space provided.

_____ **6.** The warm winds called chinooks are found on the:

a. eastern slopes of the Rocky Mountains.
b. eastern slopes of the Sierras.
c. southern slopes of the Himalayas.
d. southern slopes of the Alps.

_____ **7.** When average daily temperatures are plotted on a map, the boundaries of the zones formed roughly coincide with:

a. coastlines. **b.** longitudes.
c. mountain ranges. **d.** latitudes.

_____ **8.** At the equator, the sun's rays strike the earth's surface at an angle of about:

a. 10°. **b.** 25°. **c.** 45°. **d.** 90°.

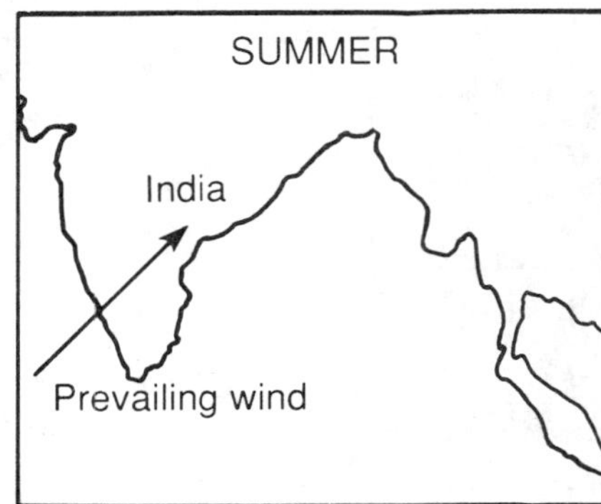

_____ **9.** What is the name for the wind in this diagram?

a. bora **b.** foehn
c. monsoon **d.** mistral

_____ **10.** Which of these statements best describes what happens to an air mass moving over a mountain range?

a. Its temperature increases moving upslope.
b. It loses moisture moving upslope.
c. Its temperature decreases moving downslope.
d. It gains moisture moving downslope.

Name ______________________ Class ______________ Date ______________

Section 26.2

Climate Zones

Choose the one best response. Write the letter of that choice in the space provided.

_____ **1.** A Mediterranean climate is characterized by winters that are:

a. mild and wet.
b. cold and wet.
c. mild and dry.
d. cold and dry.

_____ **2.** Which type of vegetation is characteristic of subarctic climates?

a. hardwood trees
b. drought-resistant shrubs
c. cone-bearing trees
d. dense grasses

_____ **3.** Rural areas usually have a lower average temperature than cities partly due to:

a. large numbers of condensation nuclei.
b. more transpiration by vegetation.
c. high atmospheric carbon dioxide levels.
d. greater available water vapor.

Read each question and answer it in the space provided.

4. What is another name for tropical grasslands? ______________

5. What type of climate is characterized by a very small temperature range and more than 250 cm of rainfall per year? ______________

6. In what type of climate was the largest yearly temperature range recorded? ______________

7. Which type of desert is characterized by cold winters and hot summers? ______________

8. What imaginary line lies in the center of the tropics? ______________

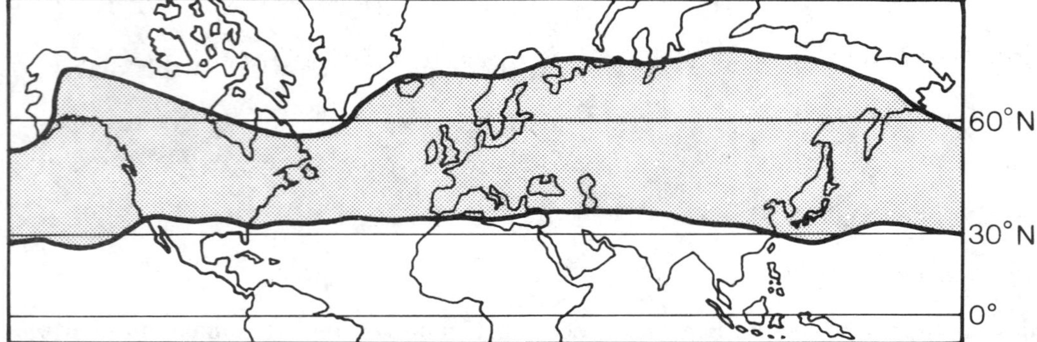

9. Which major climate zone is found in the shaded region of the diagram? ______________

10. Which type of climate do the Sahara and Kalahari deserts have? ______________

Name ______________________ Class ______________ Date ______________

MODERN EARTH SCIENCE

Section 27.1

Characteristics of Stars

Choose the one best response. Write the letter of that choice in the space provided.

_____ **1.** The apparent change in the position of an object resulting from a change in the angle or position from which it is viewed is called:

a. stellar magnitude. **b.** spectroscopy.
c. the Doppler effect. **d.** parallax.

_____ **2.** The apparent magnitude of a star is a measure of its:

a. surface temperature and composition.
b. distance from the earth.
c. brightness as it appears from earth.
d. position on the H-R diagram.

_____ **3.** A star that experiences regular changes in brightness over time is called a:

a. Cepheid variable. **b.** supernova.
c. binary star. **d.** quasar.

CATEGORY	COLOR	SURFACE TEMPERATURE (°C)
M		less than 3,500
K	orange	3,500–5,000
G		5,000–6,000
F	yellow-white	6,000–7,500
A		7,500–10,000
B	blue-white	10,000–30,000
O		above 30,000

_____ **4.** The color of a star in category O of this table would most likely be:

a. blue. **b.** blue-white.
c. yellow. **d.** red.

_____ **5.** A star in category G would most likely be:

a. blue. **b.** blue-white.
c. yellow. **d.** red.

Complete each statement by writing the correct term or phrase in the space provided.

6. From the Northern Hemisphere, circumpolar stars appear to be circling the

______________________ .

7. The apparent shift in the spectrum of a star as it moves toward or away from the earth is explained by the phenomenon known as the ______________________ .

8. The H-R diagram plots the surface temperatures of stars against their

______________________ .

9. The apparent motion of stars is due to movement of ______________________ .

10. The distance from the earth to a star that has identical apparent and absolute magnitudes is

______________________ .

Name ____________________ Class ____________ Date ____________

MODERN EARTH SCIENCE

Section 27.2

Stellar Evolution

Read each statement below. If the statement is true, write *T* in the space provided. If the statement is false, write *F* in the space provided.

_____ **1.** Nebulae never produce more than one star.

_____ **2.** A main-sequence star generates energy by fusing helium into hydrogen.

_____ **3.** A dying star can shed its outer atmosphere as a planetary nebula.

_____ **4.** The temperature inside a protostar increases due to particle collisions.

_____ **5.** As white dwarfs cool, they often explode as supernovas.

Choose the one best response. Write the letter of that choice in the space provided.

_____ **6.** In the last stage of stellar evolution following a supernova, stars too massive to form neutron stars may form a:

a. black dwarf. **b.** supergiant. **c.** white dwarf. **d.** black hole.

_____ **7.** In which stage of stellar evolution does combined hydrogen fusion and helium fusion cause a star's outer shell to expand?

a. supernova **b.** giant
c. main-sequence **d.** protostar

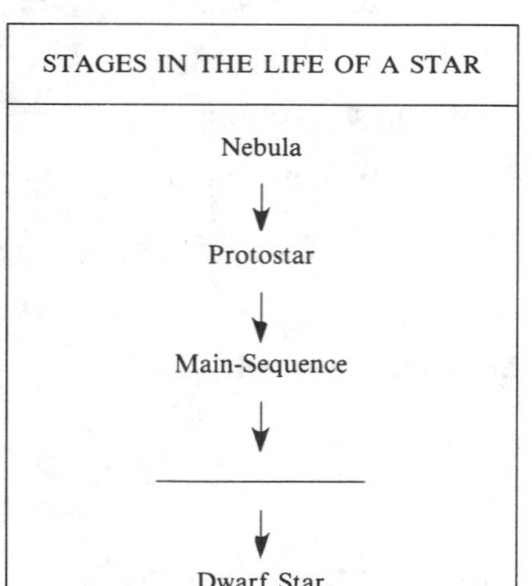

_____ **8.** What stage in the life of a star ten times more massive than the sun is missing in this chart?

a. supernova **b.** neutron star
c. black hole **d.** giant

_____ **9.** Which type of star maintains a stable size because the energy from fusion balances the force of gravity?

a. main-sequence **b.** neutron star
c. pulsar **d.** supergiant

_____ **10.** Nuclear fusion begins when temperatures within a protostar reach over:

a. 30,000°C. **b.** 300,000°C.
c. 10,000,000°C. **d.** 150,000,000°C.

Name ______________________ Class ______________ Date ______________

MODERN EARTH SCIENCE

Section 27.3

Star Groups

Read each statement below. If the statement is true, write *T* in the space provided. If the statement is false, write *F* in the space provided.

_____ **1.** A constellation is a type of galaxy.

_____ **2.** Many galaxies contain nebulae as well as stars.

_____ **3.** The brightest star in a constellation is usually labeled with the Greek beta symbol.

_____ **4.** According to the big bang theory, pulsars were among the first objects formed after the gigantic explosion that formed the universe.

_____ **5.** Stars are distributed unevenly in an irregular galaxy.

Complete each statement by writing the correct term or phrase in the space provided.

6. Pairs of stars that revolve around each other are called ______________________ .

7. Scientists continue to divide the sky into sectors using star patterns called

______________________ .

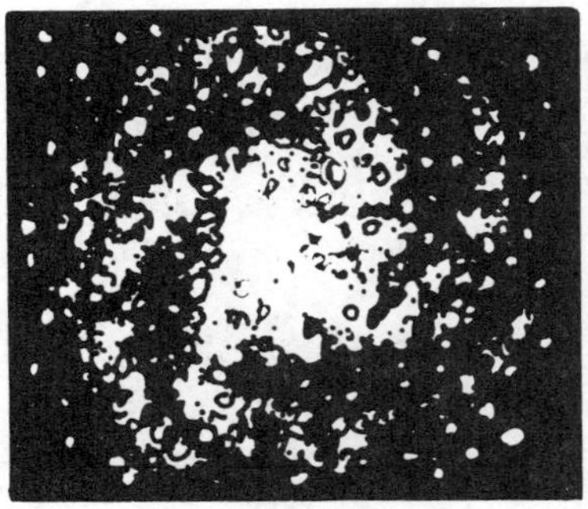

8. Galaxies such as the one shown in this diagram are known as ______________________ .

9. The starlike objects whose light must travel for 12 billion years to reach the earth are called

______________________ .

10. The big bang theory is a widely accepted theory explaining the formation of

______________________ .

Name ______________________ Class ______________ Date ______________

MODERN EARTH SCIENCE

Section 28.1

Structure of the Sun

Read each statement below. If the statement is true, write *T* in the space provided. If the statement is false, write *F* in the space provided.

_____ **1.** The core of the sun is made up of dense gas.

_____ **2.** Nuclear fusion occurs in the sun's core.

_____ **3.** The corona is part of the sun's magnetosphere.

_____ **4.** The photosphere is considered to be the surface of the sun.

_____ **5.** The layer of the sun immediately surrounding the core is called the convective zone.

Choose the one best response. Write the letter of that choice in the space provided.

_____ **6.** Ions that stream out into space through holes in the corona create the:

a. ozone layer. **b.** solar wind.
c. chromosphere. **d.** solar nebula.

_____ **7.** Which layer of the sun gives off the visible light normally seen from the earth?

a. core **b.** magnetosphere
c. corona **d.** photosphere

The Sun

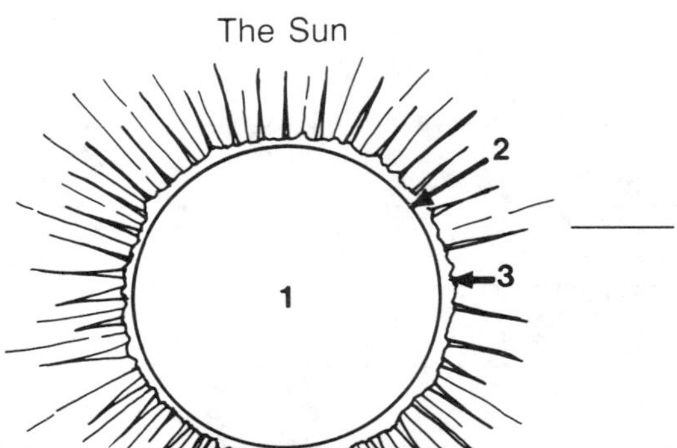

_____ **8.** Which part of the diagram represents the color sphere?

a. 1 **b. 2**
c. 3 **d. 4**

_____ **9.** The transfer of energy by moving liquids or gases occurs by the process called:

a. convection. **b.** granulation. **c.** fusion. **d.** radiation.

_____ **10.** The fusion of hydrogen nuclei in the sun produces a nucleus of:

a. hydrogen. **b.** oxygen. **c.** helium. **d.** carbon.

Name ______________________ Class ______________ Date ______________

Section 28.2

Solar Activity

Read each statement below. If the statement is true, write *T* in the space provided. If the statement is false, write *F* in the space provided.

_____ **1.** Solar flares usually occur near sunspots.

_____ **2.** Prominences form arches from one sunspot area to another.

_____ **3.** Bands of colored light that result from a magnetic storm are called solar flares.

_____ **4.** The number of sunspots increases at the beginning of a sunspot cycle.

_____ **5.** Prominences are the most violent of all solar disturbances.

_____ **6.** Auroras are usually seen close to the earth's magnetic poles.

_____ **7.** Magnetic storms can disrupt radio communications on earth.

_____ **8.** The sunspot cycle lasts an average of 20 years.

_____ **9.** The movement of sunspots was one of the first indications to astronomers that the sun rotates on its axis.

Read the statement and answer it in the space provided.

10. Explain how sunspots are formed.

____

____

____

____

____

____

Name ______________________ Class ______________ Date ______________

MODERN EARTH SCIENCE

Section 28.3

Formation of the Solar System

Choose the one best response. Write the letter of that choice in the space provided.

_____ 1. The process of photosynthesis increases the amount of which gas in the earth's atmosphere?

a. nitrogen
b. carbon dioxide
c. oxygen
d. water vapor

_____ 2. The inner planets may have lost their original atmospheres of lighter gases because of an intense solar:

a. wind. b. flare. c. prominence. d. nebula.

_____ 3. Which of the following planets originated from one of the four protoplanets closest to the sun?

a. Uranus b. Mars c. Saturn d. Pluto

Major Components of the Earth's Atmospheres

First Atmosphere	Present Atmosphere
hydrogen _______	nitrogen oxygen

_____ 4. Which of the following gases is indicated by the blank line in this table?

a. helium
b. ammonia
c. methane
d. carbon dioxide

_____ 5. Which of the following is the approximate percentage of the matter in the solar nebula that became part of the sun?

a. 38% b. 55% c. 73% d. 99%

Complete each statement by writing the correct term or phrase in the space provided.

6. The hypothesis that the sun and planets formed out of the same cloud of dust and gas was proposed by a mathematician named ______________________.

7. As the earth's atmosphere developed, much of the carbon dioxide that was present in the atmosphere was absorbed by the ______________________.

8. Astronomers have described Pluto as an ice ball of frozen gases and ______________________.

9. The large outer planets that retained much of their original atmospheres are called ______________________.

10. The small bodies of matter that eventually formed protoplanets are called ______________________.

Name ______________________ Class ______________ Date ______________

MODERN EARTH SCIENCE

Section 29.1

Models of the Solar System

Read each statement below. If the statement is true, write *T* in the space provided. If the statement is false, write *F* in the space provided.

_____ **1.** Ptolemy's model of the universe proposed that each planet has two motions.

_____ **2.** The geocentric model of the solar system states that the planets revolve around the sun.

_____ **3.** Galileo's observations confirmed the validity of a heliocentric solar system model.

_____ **4.** Kepler suggested that planets move at different speeds at different points in their orbits.

_____ **5.** According to Kepler's third law, the cube of the average distance of a planet from the sun is always proportional to the square of the planet's period.

Choose the one best response. Write the letter of that choice in the space provided.

_____ **6.** Kepler's first law states that planets orbit the sun in paths called:

a. circles. **b.** periods. **c.** epicycles. **d.** ellipses.

_____ **7.** The tendency of an object to remain at rest until acted upon by an outside force is called:

a. inertia. **b.** magnetism. **c.** gravity. **d.** velocity

_____ **8.** Based on Kepler's third law, the orbit period of an object four astronomical units from the sun would be:

a. 4 years. **b.** 8 years. **c.** 16 years. **d.** 32 years.

_____ **9.** Kepler's laws were developed from observations made by:

a. Copernicus. **b.** Galileo. **c.** Ptolemy. **d.** Brahe.

2
3 4• Sun 1

_____ **10.** In the diagram, perihelion is located at the point labeled:

a. 1. **b.** 2.
c. 3. **d.** 4.

Name ______________________ Class ______________ Date ____________

MODERN EARTH SCIENCE

Section 29.2

The Inner Planets

Choose the one best response. Write the letter of that choice in the space provided.

_____ **1.** The atmosphere of Venus is 96 percent:

a. sulfuric acid. **b.** oxygen.
c. carbon dioxide. **d.** nitrogen.

_____ **2.** The presence of a magnetic field indicates that a planet may have:

a. an iron core. **b.** an atmosphere.
c. oceans. **d.** a high surface temperature.

_____ **3.** Earth is the only planet in the solar system that has:

a. clouds. **b.** oceans of water.
c. a core. **d.** an atmosphere.

_____ **4.** Great cracks in the crust of Mars were probably formed by:

a. erosion. **b.** ice caps. **c.** meteors. **d.** fault zones.

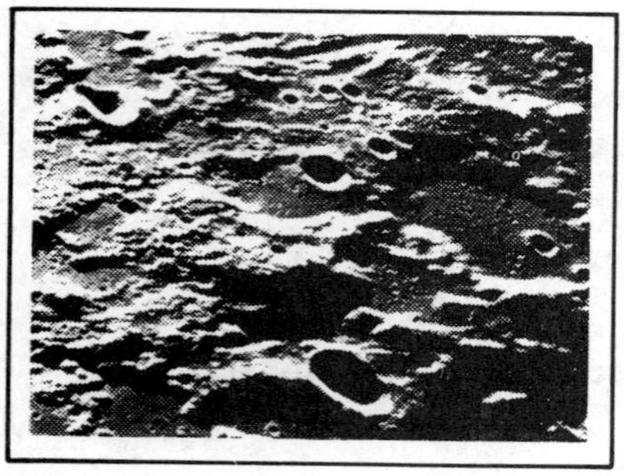

_____ **5.** In this photograph, the circular features on the planet's surface are:

a. ocean basins. **b.** valleys.
c. impact craters. **d.** volcanoes.

Complete each statement by writing the correct term or phrase in the space provided.

6. The inner planets are sometimes called ______________________ .

7. The name of the planet closest to the sun is ______________________ .

8. Almost all of the visible water on Mars appears to be trapped in ______________________ .

9. The major evidence of geologic activity on Mars is the presence of

______________________ .

10. The third planet from the sun is ______________________ .

Name ______________________ Class ______________ Date ______________

Section 29.3

The Outer Planets

Read each statement below. If the statement is true, write *T* in the space provided. If the statement is false, write *F* in the space provided.

_____ **1.** Many scientists believe that Pluto was a moon of Uranus.

_____ **2.** Due to high temperature and pressure, the surface of Jupiter is partially molten rock.

_____ **3.** Astronomers believe that the Great Red Spot is a giant atmospheric storm.

_____ **4.** The clouds covering Uranus are a greenish color.

_____ **5.** Data from the Voyager spacecraft indicate that the planet Neptune is mostly solid rock.

Complete each statement by writing the correct term or phrase in the space provided.

6. The first planet whose existence was predicted before it was discovered is

______________________ .

7. The least dense planet in the solar system is ______________________ .

8. The most distinguishing feature of Saturn is its complex system of

______________________ .

9. The planet whose axis has the greatest tilt is ______________________ .

10. The name of the planet labeled **X** in the diagram is

______________________ .

Orbits of the Outer Planets

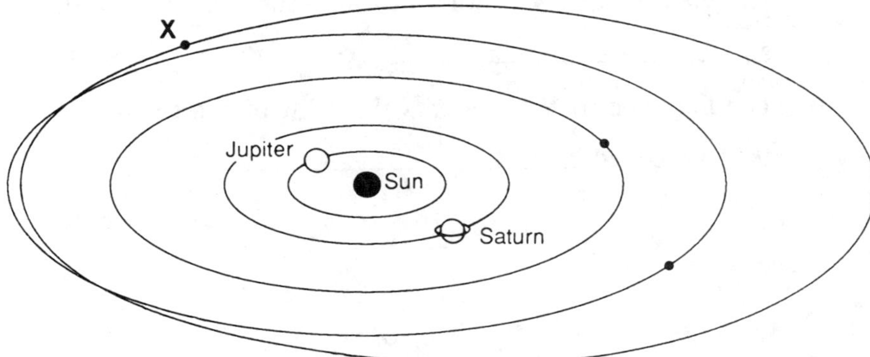

Name ______________________ Class ______________ Date ______________

Section 29.4

Asteroids, Comets, and Meteoroids

Read each statement below. If the statement is true, write *T* in the space provided. If the statement is false, write *F* in the space provided.

_____ **1.** Regardless of the direction in which a comet is traveling, its tail points away from the sun.

_____ **2.** All asteroids have the same composition.

_____ **3.** A fireball indicates that a meteorite is about to land somewhere on earth.

_____ **4.** Meteor showers occur when the earth's orbit intersects the orbits of comets.

_____ **5.** The trail of a comet can extend millions of kilometers behind it.

Complete each statement by writing the correct term or phrase in the space provided.

6. Small bits of rock or metal moving through the solar system are called

______________________ .

7. Generally, the type of meteorite that is easiest to find, due to its metallic appearance, is

______________________ .

8. The spherical cloud of gas and dust surrounding the nucleus of a comet is called

______________________ .

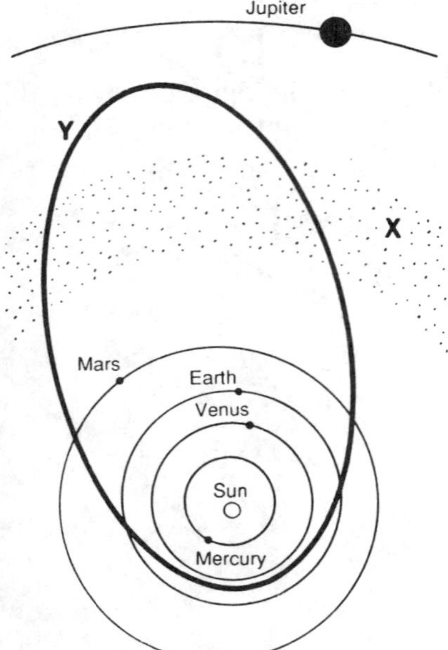

9. The area labeled **X** on the diagram represents the location of ______________________ .

10. The line labeled **Y** represents the orbit of a type of asteroid called ______________________ .

Name ______________________ Class ______________ Date ______________

Section 30.1

The Earth's Moon

Choose the one best response. Write the letter of that choice in the space provided.

_____ **1.** The earth's first artificial satellite was called:

a. *Sputnik 1.* **b.** *Explorer 1.* **c.** *Voyager 1.* **d.** *Apollo 13*

_____ **2.** A long, deep channel on the moon is called a:

a. mare. **b.** ray. **c.** streak. **d.** rille.

_____ **3.** What are bowl-shaped lunar depressions called?

a. breccia **b.** craters **c.** anorthosites **d.** basalts

_____ **4.** The outer layer of dust and small rock fragments on the moon is called:

a. maria. **b.** umbra. **c.** rille. **d.** regolith.

_____ **5.** The moon may have a small liquid core made of:

a. silicon. **b.** magnesium. **c.** iron. **d.** aluminum.

_____ **6.** During which stage of the moon's development was the surface covered with hot, molten rock?

a. first **b.** second **c.** third **d.** last

_____ **7.** The light-colored, coarse-grained rocks from the lunar highlands are called:

a. anorthosites. **b.** breccia. **c.** basalts. **d.** maria.

_____ **8.** In this diagram, the moon's mantle is the structure labeled:

a. 1. **b.** 2.
c. 3. **d.** 4.

_____ **9.** According to the giant-impact hypothesis, the moon was once part of:

a. Mercury. **b.** Venus. **c.** Earth. **d.** Mars.

Read the statement and answer it in the space provided.

10. Describe the stage of the moon's development that began when its surface cooled.

____

____

____

Name ______________________ Class ______________ Date ______________

Section 30.2

Movements of the Moon

Read each statement below. If the statement is true, write *T* in the space provided. If the statement is false, write *F* in the space provided.

______ **1.** The center of mass of the earth-moon system follows a smooth orbit around the sun.

______ **2.** One side of the moon always faces the earth.

______ **3.** The moon passes closest to the earth at apogee.

______ **4.** In the umbra, sunlight is partially blocked.

______ **5.** Lunar eclipses are visible from any location on the dark side of the earth.

Choose the one best response. Write the letter of that choice in the space provided.

______ **6.** The orbit of the moon around the earth forms:

a. a sphere. **b.** a cone. **c.** a circle. **d.** an ellipse.

______ **7.** What is the difference, in minutes, in the rising time of the moon each day?

a. 10 **b.** 25 **c.** 50 **d.** 90

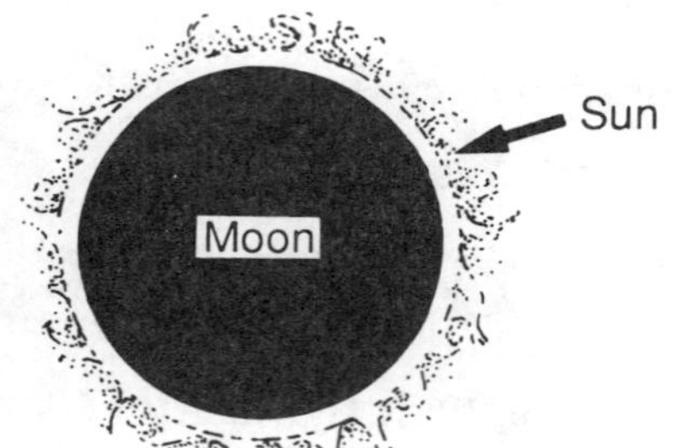

______ **8.** What type of eclipse is pictured in the diagram?

a. total solar eclipse
b. annular eclipse
c. penumbral eclipse
d. lunar eclipse

______ **9.** A total solar eclipse lasts no more than seven minutes at any location on earth because:

a. seven minutes is the time it takes for the moon to pass through the earth's penumbra.
b. the earth's rotation causes the area under the shadow of the moon to move rapidly.
c. seven minutes is the time it takes for the moon to pass through the earth's umbra.
d. the moon's spin causes its shadow to move quickly over the earth.

______ **10.** The center of mass of the earth-moon system is at a balance point, which is located:

a. within the earth's interior.
b. less than half the distance from the earth to the moon.
c. within the moon's interior.
d. more than half the distance from the earth to the moon.

Name ______________________ Class ______________ Date ______________

MODERN EARTH SCIENCE

Section 30.3

The Lunar Cycle

Choose the one best response. Write the letter of that choice in the space provided.

_____ **1.** When only a small part of the moon is visible, the moon may be in its:

a. first-quarter phase. **b.** new moon phase.
c. waning-crescent phase. **d.** last-quarter phase.

_____ **2.** Approximately how many days does it take the moon to go through a complete cycle of phases?

a. 7 **b.** 11 **c.** 27 **d.** 29

_____ **3.** Systems of measuring the passage of time are called:

a. phases. **b.** calendars. **c.** years. **d.** sols.

_____ **4.** Prior to the introduction of the Julian calendar, how many days made up a Roman year?

a. 106 **b.** 197 **c.** 265 **d.** 304

_____ **5.** Which diagram below shows the moon in a gibbous phase?

a. **b.** **c.** **d.**

Read each question and answer it in the space provided.

6. What is the major source of the light that reflects off the moon? ______________

7. What term is used to describe a decrease in the visible portion of the moon? ______________

8. What is the time required for the earth to make one complete rotation on its axis? ______________

9. What are years with an extra day called? ______________

10. What is the name of the proposed calendar that would add a day at the end of June every four years? ______________

Name ______________________ Class ______________ Date ______________

Section 30.4

Satellites of Other Planets

Read each statement below. If the statement is true, write *T* in the space provided. If the statement is false, write *F* in the space provided.

_____ **1.** Galileo discovered the four moons closest to Jupiter and what later were discovered to be the rings of Saturn.

_____ **2.** Each of the moons of Mars is much larger than the earth's moon.

_____ **3.** The two moons of Mars are believed to be fairly young because their surfaces are characterized by very few craters.

_____ **4.** Jupiter's moon Io has active volcanoes.

_____ **5.** The moon Charon is always above the same spot on the surface of Pluto because the period of Charon's orbit is equal to one day on Pluto.

_____ **6.** One hypothesis about the formation of Saturn's rings states that the rings formed when a body orbiting Saturn was torn apart by the gravity of the planet.

_____ **7.** Jupiter's moon Ganymede is the smallest and densest moon in the solar system.

Read each question and answer it in the space provided.

8. Which of Jupiter's moons may be the most densely cratered moon in the solar system? ______________________

9. What materials compose most of Saturn's rings and moons? ______________________

10. Which moon of Neptune is unusual because it travels from east to west as it revolves around the planet? ______________________

Answers to Section 1.1 Test

1. T 2. F 3. F 4. T 5. F 6. F 7. c 8. a 9. a 10. b

Guide to Tested Objectives

Objective **1**: questions 1(A), 2(A), 6(A), 9(A); Objective **2**: questions 3(A), 4(A), 5(A), 7(B), 8(A), 10(B)

Answers to Section 1.2 Test

1. T
2. F
3. T
4. F
5. T
6. meteorite
7. measuring
8. variable
9. dinosaurs
10. 14°C

Guide to Tested Objectives

Objective **3**: questions 2(B), 3(A), 5(A), 7(A), 8(B), 10(A); Objective **4**: questions 1(A), 4(A), 6(A), 9(A)

Answers to Section 1.3 Test

1. law
2. spectrum
3. Doppler effect
4. bright-line spectrum
5. spectroscope
6. expanding
7. c
8. d
9. a
10. a

Guide to Tested Objectives

Objective **5**: question 1(A); Objective **6**: questions 2(A), 3(A), 4(A), 8(B), 10(B); Objective **7**: questions 5(A), 9(A); Objective **8**: questions 6(B), 7(A)

Answers to Section 2.1 Test

1. T	2. F	3. F	4. T	5. T	6. c	7. a	8. d	9. c	10. d

Guide to Tested Objectives

Objective **1**: questions 1(A), 6(A), 7(A); Objective **2**: questions 2(A), 8(A), 9(B); Objective **3**: questions 3(A), 4(A); Objective **4**: questions 5(B), 10(A)

Answers to Section 2.2 Test

1. F	2. T	3. T	4. F	5. F	6. T	7. b	8. c	9. b	10. d

Guide to Tested Objectives

Objective **5**: questions 1(A), 4(A), 5(A); Objective **6**: questions 2(A), 8(A), 9(B), 10(B); Objective **7**: questions 3(A), 6(A), 7(A)

Answers to Section 2.3 Test

1. F	2. T	3. F	4. T	5. F	6. F	7. c	8. d	9. d	10. a

Guide to Tested Objectives

Objective **8**: questions 1(A), 3(A), 5(A), 6(B), 7(A), 9(B); Objective **9**: questions 2(A), 4(A), 8(A), 10(A)

Answers to Section 3.1 Test

1. a	2. c	3. b	4. d
5. a	6. geographic pole	7. great circle	8. prime meridian
9. North Pole	10. minutes		

Guide to Tested Objectives

Objective **1**: questions 3(A), 4(A); Objective **2**: questions 1(B), 2(B), 7(A), 8(A), 10(A); Objective **3**: questions 5(A), 6(A), 9(A)

Answers to Section 3.2 Test

1. b 2. c 3. a 4. a 5. T 6. F 7. F 8. T 9. T 10. F

Guide to Tested Objectives

Objective **4**: questions 1(A), 2(A), 3(A), 4(A), 5(A), 6(A); Objective **5**: questions 7(B), 8(A), 9(B), 10(A)

Answers to Section 3.3 Test

1. T 2. F 3. T 4. T 5. F 6. d 7. b 8. b 9. c 10. d

Guide to Tested Objectives

Objective **6**: questions 1(A), 2(A), 7(A), 8(B); Objective **7**: questions 3(A), 4(A), 5(B), 6(B), 9(A), 10(A)

Answers to Section 4.1 Test

1. T 2. T 3. F 4. T 5. F 6. T 7. d 8. c 9. b 10. c

Guide to Tested Objectives

Objective **1**: questions 2(A), 7(A); Objective **2**: questions 5(B), 8(B), 10(B); Objective **3**: questions 1(A), 3(A), 4(A), 6(A), 9(A)

Answers to Section 4.2 Test

1. a
2. d
3. a
4. c
5. transform fault
6. convection currents
7. divergent
8. continental crust
9. asthenosphere
10. plate tectonics

Guide to Tested Objectives

Objective **4**: questions 1(A), 10(B); Objective **5**: questions 2(A), 3(A), 4(A), 8(A), 9(B); Objective **6**: questions 5(A), 6(A); Objective **7**: question 7(A)

Answers to Section 5.1 Test

1. F   2. T   3. F   4. F   5. T   6. T   7. c   8. b   9. d   10. c

Guide to Tested Objectives

Objective **1**: questions 1(B), 4(A), 6(A), 7(A), 10(A); Objective **2**: questions 2(A), 3(A), 5(A), 8(B), 9(A)

Answers to Section 5.2 Test

1. T
2. F
3. T
4. F
5. F
6. plane
7. footwall
8. anticline
9. folding
10. synclines

Guide to Tested Objectives

Objective **3**: questions 1(A), 2(A), 8(B), 9(A), 10(A); Objective **4**: questions 3(B), 4(A), 5(B), 6(A), 7(A)

Answers to Section 5.3 Test

1. F   2. T   3. T   4. F   5. T   6. T   7. a   8. d   9. c   10. b

Guide to Tested Objectives

Objective **5**: questions 2(A), 3(B), 6(A), 10(B); Objective **6**: questions 1(A), 4(A), 5(A), 7(A), 8(B), 9(A)

Answers to Section 6.1 Test

1. c
2. a
3. b
4. d
5. a
6. point B
7. point D
8. fault zone
9. aftershocks
10. seismic waves

Guide to Tested Objectives

Objective **1**: questions 1(A), 2(A), 3(B), 5(B), 6(B), 7(A), 9(A), 10(A); Objective **2**: questions 4(B), 8(A)

Answers to Section 6.2 Test

1. F
2. T
3. T
4. T
5. F
6. Mercalli scale
7. microquake
8. L (or surface) waves
9. S (or secondary) waves
10. magnitude

Guide to Tested Objectives

Objective **3**: questions 4(A), 8(B), 9(B); Objective **4**: question 3(B); Objective **5**: questions 1(A), 2(A), 5(A), 6(A), 7(A), 10(A)

Answers to Section 6.3 Test

1. T
2. F
3. T
4. T
5. F
6. c
7. a
8. d
9. b
10. a

Guide to Tested Objectives

Objective **6**: questions 2(B), 5(B); Objective **7**: questions 1(A), 7(B), 10(A); Objective **8**: questions 4(A), 6(A); Objective **9**: questions 3(A), 8(B), 9(A)

Answers to Section 7.1 Test

1. F 2. T 3. F 4. b 5. a 6. c 7. d 8. b 9. a 10. a

Guide to Tested Objectives

Objective **1**: questions 1(A), 4(B), 6(A); Objective **2**: questions 2(A), 5(A), 7(A); Objective **3**: questions 3(A), 8(B), 9(A), 10(A)

Answers to Section 7.2 Test

1. b 2. c 3. a 4. c
5. d 6. b 7. lapilli 8. ash
9. volcanic bomb 10. caldera

Guide to Tested Objectives

Objective **4**: questions 2(A), 3(A); Objective **5**: questions 8(B), 9(A), 10(A); Objective **6**: questions 1(B), 4(A), 5(A); Objective **7**: questions 6(A), 7(A)

Answers to Section 7.3 Test

1. F 2. T 3. F 4. a 5. d 6. c 7. b 8. c 9. a 10. d

Guide to Tested Objectives

Objective **8**: questions 1(A), 2(A), 4(A), 5(B), 10(A); Objective **9**: questions 3(A), 6(A), 7(A), 8(A), 9(A)

Answers to Section 8.1 Test

1. F 2. T 3. T 4. F 5. T 6. d 7. b 8. d 9. a 10. b

Guide to Tested Objectives

Objective **1**: questions 1(A), 2(A); Objective **2**: questions 6(A), 8(A); Objective **3**: questions 7(B), 9(B), 10(B); Objective **4**: question 4(A); Objective **5**: questions 3(A), 5(A)

Answers to Section 8.2 Test

1. molecule 2. chemical formula 3. chemical bonds 4. solution
5. diatomic molecules 6. b 7. d 8. d
9. b 10. a

Guide to Tested Objectives

Objective **6**: questions 1(A), 5(A), 8(B); Objective **7**: questions 3(A), 10(B); Objective **8**: questions 2(A), 9(A); Objective **9**: questions 4(A), 6(B), 7(A)

Answers to Section 9.1 Test

1. T 2. F 3. T 4. F 5. T 6. a 7. d 8. c 9. c 10. d

Guide to Tested Objectives

Objective **1**: questions 4(A), 7(A); Objective **2**: questions 1(A), 3(B); Objective **3**: questions 5(A), 8(A); Objective **4**: questions 2(A), 6(A), 9(A), 10(B)

Answers to Section 9.2 Test

1. T 2. F 3. T 4. T 5. F 6. c 7. a 8. d 9. b 10. b

Guide to Tested Objectives

Objective **5**: questions 2(A), 4(A), 6(A), 9(B), 10(B); Objective **6**: questions 1(A), 3(A), 5(A), 7(A), 8(A)

Answers to Section 10.1 Test

1. F	2. T	3. T	4. F
5. T	6. cycle	7. igneous	8. pressure
9. sediment	10. lava		

Guide to Tested Objectives

Objective **1**: questions 1(A), 3(A), 4(A), 9(A), 10(A); Objective **2**: questions 2(A), 5(A), 6(A), 7(B), 8(B)

Answers to Section 10.2 Test

1. F 2. T 3. F 4. F 5. T 6. a 7. a 8. b 9. d 10. b

Guide to Tested Objectives

Objective **3**: question 5(A); Objective **4**: questions 1(A), 6(A), 9(A), 10(A); Objective **5**: questions 2(A), 3(A), 4(A), 7(B), 8(B)

Answers to Section 10.3 Test

1. T	2. F	3. T	4. T
5. F	6. compaction	7. evaporites	8. geode
9. concretions	10. chemical sedimentary rock		

Guide to Tested Objectives

Objective **6**: questions 1(A), 2(A), 4(A), 5(A), 6(A), 7(A), 10(A); Objective **7**: questions 3(A), 8(B), 9(B)

Answers to Section 10.4 Test

1. F	2. T	3. T	4. T
5. T	6. sandstone	7. contact metamorphism	8. foliated
9. shale	10. regional metamorphism		

Guide to Tested Objectives

Objective **8**: questions 1(A), 2(B), 3(A), 4(A), 5(A), 6(B), 7(A), 10(A); Objective **9**: questions 8(A), 9(A)

Answers to Section 11.1 Test

1. F 2. T 3. F 4. a 5. b 6. a 7. d 8. a 9. b 10. a

Guide to Tested Objectives

Objective **1**: questions 1(A), 5(A), 6(B), 8(A), 9(A); Objective **2**: questions 2(A), 3(A), 4(A), 7(B), 10(A)

Answers to Section 11.2 Test

1. F 2. T 3. T 4. c 5. b 6. a 7. b 8. b 9. d 10. a

Guide to Tested Objectives

Objective **3**: questions 2(A), 4(A), 5(B); Objective **4**: questions 6(A), 7(B); Objective **5**: questions 1(A), 8(A); Objective **6**: question 9(B); Objective **7**: questions 3(A), 10(B)

Answers to Section 11.3 Test

1. T 2. F 3. T 4. d 5. b 6. c 7. d 8. a 9. b 10. c

Guide to Tested Objectives

Objective **8**: questions 1(A), 2(B), 4(A), 5(A), 6(B), 9(B), 10(A); Objective **9**: questions 3(A), 7(B), 8(A)

Answers to Section 11.4 Test

1. d 2. b 3. a 4. c
5. a 6. b 7. solar 8. solar collector
9. geothermal energy 10. active and passive

Guide to Tested Objectives

Objective **10**: questions 7(A), 8(A), 10(A); Objective **11**: questions 5(A), 6(B), 9(B); Objective **12**: questions 1(A), 2(A), 3(A), 4(A)

Answers to Section 12.1 Test

1. T 2. F 3. T 4. F 5. F 6. c 7. d 8. a 9. c 10. d

Guide to Tested Objectives

Objective **1**: questions 2(A), 3(A), 5(A), 6(B), 8(A); Objective **2**: questions 1(A), 4(A), 7(A), 9(B)

Answers to Section 12.2 Test

1. a
2. c
3. b
4. c
5. topography
6. surface area
7. ice wedging
8. two
9. carbonation
10. calcium bicarbonate

Guide to Tested Objectives

Objective **3**: questions 1(A), 9(B); Objective **4**: questions 6(A), 7(A), 8(B); Objective **5**: questions 2(B), 3(A), 4(A), 5(A), 10(A)

Answers to Section 12.3 Test

1. T
2. F
3. T
4. T
5. T
6. laterite
7. humus
8. horizons
9. A
10. D

Guide to Tested Objectives

Objective **6**: questions 2(A), 5(A), 7(A), 8(A), 9(B), 10(B); Objective **7**: questions 1(A), 3(A), 4(A), 6(A)

Answers to Section 12.4 Test

1. T 2. F 3. F 4. T 5. T 6. d 7. a 8. b 9. b 10. a

Guide to Tested Objectives

Objective **8**: questions 3(A), 5(A); Objective **9**: questions 4(A), 6(A), 9(A); Objective **10**: questions 1(A), 2(A); Objective **11**: questions 7(A), 8(B), 10(A)

Answers to Section 13.1 Test

1. T 2. F 3. T 4. T 5. F 6. T 7. a 8. b 9. b 10. c

Guide to Tested Objectives

Objective **1**: questions 1(A), 2(A), 3(A), 9(A), 10(B); Objective **2**: questions 4(A), 6(A), 8(B); Objective **3**: questions 5(A), 7(A)

Answers to Section 13.2 Test

1. a
2. c
3. c
4. c
5. a
6. evapotranspiration
7. watersheds
8. stream piracy
9. headward erosion
10. runoff

Guide to Tested Objectives

Objective **4**: questions 1(A), 2(A), 6(A), 7(A), 8(A), 9(B), 10(A); Objective **5**: questions 3(A), 4(B); Objective **6**: question 5(A)

Answers to Section 13.3 Test

1. F
2. T
3. F
4. F
5. T
6. T
7. T
8. deposition
9. lake
10. levees

Guide to Tested Objectives

Objective **7**: questions 1(B), 2(A), 8(B); Objective **8**: questions 3(A), 5(A); Objective **9**: questions 4(A), 6(A), 7(A), 9(A), 10(A)

Answers to Section 14.1 Test

1. b
2. a
3. d
4. b
5. a
6. b
7. groundwater
8. gradient
9. permeability
10. an aquifer

Guide to Tested Objectives

Objective **1**: questions 1(B), 5(A), 9(A); Objective **2**: questions 3(A), 6(B), 7(A), 10(A); Objective **3**: questions 2(A), 8(A); Objective **4**: question 4(A)

Answers to Section 14.2 Test

1. F
2. T
3. T
4. T
5. T
6. c
7. d
8. b
9. d
10. a

Guide to Tested Objectives

Objective **2**: question 1(A); Objective **5**: questions 2(A), 5(A), 6(A), 7(A), 10(A); Objective **6**: questions 3(A), 4(A), 8(B), 9(B)

Answers to Section 14.3 Test

1. c
2. a
3. d
4. c
5. sinkhole
6. chemical weathering
7. soft water
8. karst topography
9. caverns
10. natural bridge

Guide to Tested Objectives

Objective **7**: questions 2(A), 3(A), 5(A), 6(B), 7(A), 9(A), 10(A); Objective **8**: questions 1(A), 4(B), 8(A)

Answers to Section 15.1 Test

1. d
2. b
3. c
4. d
5. a
6. pressure
7. thickness
8. glaciers
9. snowline
10. Antarctica

Guide to Tested Objectives

Objective **1**: questions 6(A), 8(A), 9(A); Objective **2**: question 10(A); Objective **3**: questions 4(A), 5(B), 7(A); Objective **4**: question 1(A); Objective **5**: question 2(A); Objective **6**: question 3(A)

Answers to Section 15.2 Test

1. T
2. F
3. F
4. T
5. F
6. T
7. 2
8. 4
9. polishing
10. precipitation

Guide to Tested Objectives

Objective **2**: question 1(A); Objective **4**: questions 5(A), 9(B); Objective **5**: questions 2(A), 3(A), 6(A), 7(A), 8(A); Objective **6**: questions 4(B), 10(A)

Answers to Section 15.3 Test

1. F
2. T
3. T
4. F
5. F
6. F
7. interglacial periods
8. B
9. Milankovitch
10. ice age

Guide to Tested Objectives

Objective **7**: questions 1(A), 2(A), 4(A), 5(B), 6(B), 7(A), 10(A); Objective **8**: questions 3(A), 8(B), 9(A)

Answers to Section 16.1 Test

1. F 2. F 3. T 4. T 5. T 6. b 7. a 8. b 9. c 10. d

Guide to Tested Objectives

Objective **1**: questions 1(A), 2(A), 6(A), 7(A); Objective **2**: questions 3(A), 4(A), 5(A), 8(A), 9(A), 10(B)

Answers to Section 16.2 Test

1. d
2. a
3. b
4. c
5. c
6. sea arch
7. wave-built terraces
8. sand bars
9. longshore
10. east

Guide to Tested Objectives

Objective **3**: questions 1(A), 2(A), 6(A), 7(A); Objective **4**: questions 3(A), 8(A); Objective **5**: questions 4(A), 5(A), 9(B), 10(B)

Answers to Section 16.3 Test

1. a
2. b
3. b
4. d
5. d
6. c
7. c
8. tidal flats
9. atoll
10. fiords

Guide to Tested Objectives

Objective **6**: questions 1(A), 2(A), 10(A); Objective **7**: questions 4(B), 5(B), 8(A), 9(A); Objective **8**: questions 6(A), 7(A); Objective **9**: question 3(A)

Answers to Section 17.1 Test

1. T
2. F
3. F
4. T
5. T
6. strata
7. ripple marks
8. bedding plane
9. law of superposition
10. an angular unconformity

Guide to Tested Objectives

Objective **1**: question 1(A); Objective **2**: questions 6(A), 7(A), 8(A), 9(A); Objective **3**: questions 2(B), 4(A), 5(A), 10(B); Objective **4**: question 3(A)

Answers to Section 17.2 Test

1. a
2. b
3. d
4. a
5. c
6. 39
7. carbon-14
8. protons and neutrons
9. erosion rates
10. 30 cm per 1,000 years

Guide to Tested Objectives

Objective **5**: questions 9(A), 10(A); Objective **6**: questions 3(A), 4(A), 5(B); Objective **7**: questions 1(A), 2(B), 6(A), 7(A), 8(A)

Answers to Section 17.3 Test

1. T
2. F
3. T
4. T
5. F
6. b
7. a
8. c
9. b
10. b

Guide to Tested Objectives

Objective **8**: questions 2(A), 4(A), 6(A); Objective **9**: questions 3(A), 5(A), 7(A), 8(A); Objective **10**: questions 1(A), 9(B), 10(B)

Answers to Section 18.1 Test

1. T
2. T
3. F
4. F
5. T
6. fossil content
7. Paleozoic Era
8. layer **4**
9. in the ocean
10. Cenozoic Era

Guide to Tested Objectives

Objective **1**: questions 1(A), 2(B), 4(A), 6(A), 8(B); Objective **2**: questions 3(A), 5(A), 7(A), 9(A), 10(A)

Answers to Section 18.2 Test

1. c
2. d
3. b
4. d
5. a
6. Cenozoic
7. Pliocene
8. Silurian
9. stromatolites
10. Tertiary

Guide to Tested Objectives

Objective **3**: questions 1(A), 2(A), 9(B); Objective **4**: questions 3(B), 4(A), 8(B); Objective **5**: question 5(A); Objective **6**: questions 6(A), 7(A), 10(A)

Answers to Section 19.1 Test

1. d
2. c
3. a
4. c
5. a
6. Eurasia
7. plate tectonics
8. Alfred Wegener
9. Gondwanaland
10. Tethys Sea

Guide to Tested Objectives

Objective **1**: questions 2(A), 3(A), 7(A); Objective **2**: questions 1(A), 4(B), 5(B), 6(A), 8(A), 9(A), 10(A)

Answers to Section 19.2 Test

1. F
2. F
3. T
4. T
5. T
6. a
7. c
8. b
9. c
10. d

Guide to Tested Objectives

Objective **3**: questions 1(A), 2(A), 3(A), 4(A), 5(A), 7(A), 8(B); Objective **4**: questions 6(A), 9(B), 10(A)

Answers to Section 19.3 Test

1. T
2. F
3. F
4. F
5. T
6. Colorado River
7. Colorado Plateau
8. erosion
9. Cambrian Period
10. shale

Guide to Tested Objectives

Objective **5**: questions 1(A), 2(A), 3(A), 6(A), 7(A), 8(A), 9(B), 10(A); Objective **6**: questions 4(A), 5(A)

Answers to Section 20.1 Test

1. T	2. T	3. T	4. F
5. T	6. Atlantic Ocean	7. submersible	8. sonar
9. Pacific Ocean	10. Indian Ocean		

Guide to Tested Objectives

Objective **1**: questions 2(A), 3(A), 6(A), 9(A), 10(B); Objective **2**: questions 1(A), 4(B), 5(A), 7(A), 8(A)

Answers to Section 20.2 Test

1. c	2. b	3. a	4. d
5. b	6. deep ocean basin	7. submarine canyons	8. abyssal plain
9. continental shelf	10. base of the continental slope		

Guide to Tested Objectives

Objective **3**: questions 1(A), 2(A), 3(A), 7(A), 9(A), 10(B); Objective **4**: questions 4(A), 5(B), 6(A), 8(A)

Answers to Section 20.3 Test

1. F	2. T	3. F	4. T
5. F	6. F	7. siliceous ooze	8. D
9. nodules	10. mud		

Guide to Tested Objectives

Objective **5**: questions 1(A), 4(A), 5(A), 6(A), 8(B); Objective **6**: questions 2(A), 3(A), 7(A), 9(A), 10(B)

Answers to Section 21.1 Test

1. F	2. T	3. F	4. T
5. T	6. carbon dioxide	7. thermocline	8. pack ice
9. chlorine	10. reflected		

Guide to Tested Objectives

Objective **1**: questions 1(B), 2(A), 5(A), 6(A), 9(B); Objective **2**: questions 3(A), 4(A), 7(A), 8(A), 10(A)

Answers to Section 21.2 Test

1. a	2. d	3. b	4. d
5. a	6. plankton	7. photosynthesis	8. benthos
9. neritic zone	10. intertidal zone		

Guide to Tested Objectives

Objective **3**: questions 1(A), 2(A), 8(A); Objective **4**: questions 3(B), 5(A), 6(A), 7(A); Objective **5**: questions 4(B), 9(B), 10(A)

Answers to Section 21.3 Test

1. F 2. T 3. T 4. F 5. T 6. c 7. b 8. a 9. c 10. b

Guide to Tested Objectives

Objective **6**: questions 1(A), 2(A), 4(A), 5(A), 6(B), 7(A), 8(B), 9(A); Objective **7**: questions 3(A), 10(A)

Answers to Section 22.1 Test

1. b
2. a
3. a
4. d
5. b
6. northeast
7. the Coriolis effect
8. Sargasso Sea
9. Gulf Stream
10. Antarctic Bottom Water

Guide to Tested Objectives

Objective **1**: questions 1(A), 3(A), 5(B), 6(B), 7(A), 8(A), 9(A); Objective **2**: questions 2(A), 4(A), 10(A)

Answers to Section 22.2 Test

1. F
2. T
3. T
4. F
5. T
6. b
7. a
8. c
9. c
10. a

Guide to Tested Objectives

Objective **3**: questions 2(A), 5(A), 6(A), 7(A), 10(B); Objective **4**: questions 1(A), 3(A), 4(B), 8(A), 9(B)

Answers to Section 22.3 Test

1. F
2. F
3. T
4. F
5. T
6. T
7. flood tide
8. spring tide
9. tidal bore
10. neap tide

Guide to Tested Objectives

Objective **5**: questions 1(A), 2(B), 3(B), 4(A), 6(A), 8(B), 10(A); Objective **6**: questions 5(A), 7(A), 9(A)

Answers to Section 23.1 Test

1. c
2. a
3. b
4. d
5. b
6. carbon dioxide
7. atmospheric pressure
8. troposphere
9. an air pollutant
10. weather

Guide to Tested Objectives

Objective **1**: questions 3(B), 6(A), 10(A); Objective **2**: questions 2(B), 5(A), 7(A); Objective **3**: questions 1(A), 8(A); Objective **4**: questions 4(A), 9(A)

Answers to Section 23.2 Test

1. F
2. F
3. T
4. T
5. T
6. albedo
7. scattering
8. convection
9. greenhouse effect
10. electromagnetic spectrum

Guide to Tested Objectives

Objective **5**: questions 1(B), 3(A), 6(A), 7(A), 10(A); Objective **6**: questions 4(A), 5(A), 9(A); Objective **7**: questions 2(A), 8(B)

Answers to Section 23.3 Test

1. F
2. F
3. T
4. F
5. T
6. T
7. a
8. d
9. c
10. d

Guide to Tested Objectives

Objective **8**: questions 2(A), 3(A), 4(A), 6(B), 7(A), 8(A), 9(B), 10(A); Objective **9**: questions 1(A), 5(A)

Answers to Section 24.1 Test

1. F	2. T	3. T	4. F
5. T	6. psychrometer	7. water vapor	8. 41%
9. ocean water	10. frost		

Guide to Tested Objectives

Objective **1**: questions 1(A), 5(A), 7(A), 9(A); Objective **2**: questions 4(B), 6(A), 8(B); Objective **3**: questions 2(A), 3(A), 10(A)

Answers to Section 24.2 Test

1. b	2. c	3. a	4. b
5. d	6. condensation level	7. steam fog	8. adiabatic
9. cirrostratus	10. thunderheads		

Guide to Tested Objectives

Objective **4**: questions 4(A), 5(B), 6(A), 8(A); Objective **5**: questions 1(A), 3(B), 9(A), 10(A); Objective **6**: questions 2(A), 7(A)

Answers to Section 24.3 Test

1. F	2. T	3. T	4. T	5. F	6. a	7. b	8. b	9. d	10. c

Guide to Tested Objectives

Objective **7**: questions 1(A), 5(A), 6(B), 9(A); Objective **8**: questions 3(A), 7(A), 10(A); Objective **9**: questions 2(A), 4(A); Objective **10**: question 8(B)

Answers to Section 25.1 Test

1. T 2. T 3. F 4. T 5. F 6. d 7. b 8. d 9. b 10. a

Guide to Tested Objectives

Objective 1: questions 1(A), 2(A), 5(A); Objective 2: questions 3(A), 4(A), 6(B), 7(B), 8(B), 9(A), 10(A)

Answers to Section 25.2 Test

1. thunderstorm
2. warm front
3. polar front
4. stationary
5. Coriolis effect
6. typhoons
7. tornado
8. eye
9. occluded front
10. Hurricanes form when warm, moist air over an ocean rises rapidly. The rising air is cooled and moisture in the air condenses, releasing large amounts of energy as latent heat. This heat increases the force of the rising air. More moist tropical air is drawn into the column by the rising air, releasing more heat and sustaining the process. The storm spins as the result of the Coriolis effect.

Guide to Tested Objectives

Objective 3: questions 1(A), 2(A), 9(A); Objective 4: questions 3(A), 4(B), 5(A); Objective 5: questions 6(A), 7(A), 8(A), 10(B)

Answers to Section 25.3 Test

1. d
2. d
3. b
4. a
5. d
6. Fahrenheit and Celsius
7. northerly
8. anemometer
9. wind speed and direction
10. relative humidity

Guide to Tested Objectives

Objective 6: questions 1(A), 2(A), 3(A), 6(A), 7(A); Objective 7: questions 4(A), 5(B), 8(A), 9(A), 10(B)

Answers to Section 25.4 Test

1. T
2. T
3. F
4. F
5. F
6. isobar
7. northerly (0°)
8. overcast
9. precipitation
10. 48 hours

Guide to Tested Objectives

Objective 8: questions 2(A), 3(A), 6(A), 7(B), 8(B); Objective 9: questions 1(A), 4(A), 5(A), 9(A), 10(A)

Answers to Section 26.1 Test

1. F 2. T 3. T 4. F 5. F 6. a 7. d 8. d 9. c 10. b

Guide to Tested Objectives

Objective **1**: questions 3(A), 4(B), 7(A), 8(A); Objective **2**: questions 1(A), 5(A), 9(A); Objective **3**: questions 2(A), 6(A), 10(B)

Answers to Section 26.2 Test

1. a
2. c
3. b
4. tropical savanna
5. rain forest
6. subarctic
7. middle-latitude
8. the equator
9. middle-latitude climates
10. tropical desert

Guide to Tested Objectives

Objective **4**: questions 5(B), 8(A), 9(A), 10(A); Objective **5**: questions 2(A), 6(B); Objective **6**: questions 1(A), 4(A), 7(A); Objective **7**: question 3(A)

Answers to Section 27.1 Test

1. d
2. c
3. a
4. a
5. c
6. North Star (Polaris)
7. Doppler effect
8. absolute magnitudes
9. the earth
10. 32.6 light-years

Guide to Tested Objectives

Objective **1**: questions 4(B), 5(B); Objective **2**: questions 6(A), 7(A), 9(A); Objective **3**: questions 1(A), 3(A); Objective **4**: questions 2(A), 8(A), 10(A)

Answers to Section 27.2 Test

1. F
2. F
3. T
4. T
5. F
6. d
7. b
8. d
9. a
10. c

Guide to Tested Objectives

Objective **5**: questions 1(A), 4(A), 10(B); Objective **6**: questions 2(B), 9(A); Objective **7**: questions 3(A), 5(A), 6(A), 7(A), 8(B)

Answers to Section 27.3 Test

1. F
2. T
3. F
4. F
5. T
6. binary stars
7. constellations
8. spiral (or barred-spiral) galaxies
9. quasars
10. the universe

Guide to Tested Objectives

Objective **8**: questions 1(A), 3(A), 7(A); Objective **9**: questions 2(A), 5(A), 6(A), 8(B); Objective **10**: questions 4(B), 9(A), 10(A)

Answers to Section 28.1 Test

1. T 2. T 3. F 4. T 5. F 6. b 7. d 8. c 9. a 10. c

Guide to Tested Objectives

Objective **1**: questions 1(A), 2(A), 10(A); Objective **2**: questions 5(A), 9(A); Objective **3**: questions 3(A), 4(A), 6(A), 7(B), 8(B)

Answers to Section 28.2 Test

1. T 2. T 3. F 4. T 5. F 6. T 7. T 8. F 9. T

10. The up-and-down movement of gases in the sun's convective zone and the movement of the sun's rotation produce magnetic fields that slow down activity in the convective zone. Slower convection means less gas is transferring heat from the core to the photosphere. Therefore, regions of the photosphere near these magnetic fields are cooler and appear darker than surrounding areas. These darker areas are sunspots.

Guide to Tested Objectives

Objective **4**: questions 4(A), 8(A), 9(B), 10(B); Objective **5**: questions 1(A), 2(A), 5(A), 7(A); Objective **6**: questions 3(A), 6(A)

Answers to Section 28.3 Test

1. c 2. a 3. b 4. a
5. d 6. Laplace 7. ocean water 8. rock
9. gas giants 10. planetesimals

Guide to Tested Objectives

Objective **7**: questions 5(B), 6(B), 10(A); Objective **8**: questions 2(A), 8(A), 9(A); Objective **9**: questions 1(A), 3(A), 4(A), 7(A)

Answers to Section 29.1 Test

1. T 2. F 3. T 4. T 5. T 6. d 7. a 8. b 9. d 10. c

Guide to Tested Objectives

Objective 1: questions 1(A), 2(A), 3(A), 9(A), 10(A); Objective 2: questions 4(A), 5(B), 6(A), 7(A), 8(B)

Answers to Section 29.2 Test

1. c
2. a
3. b
4. d
5. c
6. terrestrial planets
7. Mercury
8. polar ice caps
9. volcanoes
10. the earth

Guide to Tested Objectives

Objective 3: questions 1(A), 2(B), 6(A), 7(A); Objective 4: questions 3(A), 4(A), 5(A), 8(A), 9(B), 10(A)

Answers to Section 29.3 Test

1. F
2. F
3. T
4. T
5. F
6. Neptune
7. Saturn
8. rings
9. Uranus
10. Pluto

Guide to Tested Objectives

Objective 5: questions 2(B), 3(A), 7(B), 8(A); Objective 6: questions 1(A), 4(A), 5(A), 6(A), 9(A), 10(A)

Answers to Section 29.4 Test

1. T
2. F
3. F
4. T
5. T
6. meteoroids
7. iron
8. a coma
9. the asteroid belt
10. an earth-grazer

Guide to Tested Objectives

Objective 7: questions 1(A), 2(A), 8(A), 9(A), 10(B); Objective 8: questions 3(A), 4(A), 5(B), 6(A), 7(A)

Answers to Section 30.1 Test

1. a 2. d 3. b 4. d 5. c 6. b 7. a 8. b 9. c

10. The outer surface of the moon cooled to form a solid crust over the molten rock. Some meteorites punctured this crust and molten rock flowed out onto the surface through the breaks. This molten rock partially filled the craters and formed the maria.

Guide to Tested Objectives

Objective **1**: questions 1(A), 2(A), 3(A), 4(A), 7(A); Objective **2**: questions 5(A), 8(B); Objective **3**: questions 6(A), 9(A), 10(B)

Answers to Section 30.2 Test

1. T 2. T 3. F 4. F 5. T 6. d 7. c 8. b 9. b 10. a

Guide to Tested Objectives

Objective **4**: questions 1(A), 2(A), 3(A), 6(A), 7(A), 10(B); Objective **5**: questions 4(A), 5(A), 8(B), 9(B)

Answers to Section 30.3 Test

1. c
2. d
3. b
4. d
5. c
6. the sun
7. waning
8. a day
9. leap years
10. World Calendar

Guide to Tested Objectives

Objective **6**: questions 1(A), 2(A), 5(B), 6(A), 7(A); Objective **7**: questions 3(A), 4(B), 8(A), 9(A), 10(A)

Answers to Section 30.4 Test

1. T
2. F
3. F
4. T
5. T
6. T
7. F
8. Callisto
9. ice and rock
10. Triton

Guide to Tested Objectives

Objective **8**: questions 2(A), 3(A); Objective **9**: questions 1(A), 4(A), 5(A), 6(B), 7(A), 8(B), 9(A), 10(A)